The Principles of Life

The Principles of Life

Tibor Gánti

With a commentary by
James Griesemer and Eörs Szathmáry

OXFORD
UNIVERSITY PRESS

OXFORD
UNIVERSITY PRESS

Great Clarendon Street, Oxford OX2 6DP
Oxford University Press is a department
of the University of Oxford.
It furthers the University's objective of excellence
in research, scholarship, and education by
publishing worldwide in

Oxford   New York

Auckland   Bangkok   Buenos Aires
Cape Town   Chennai   Dar es Salaam
Delhi   Hong Kong   Istanbul   Karachi   Kolkata
Kuala Lumpur   Madrid   Melbourne   Mexico City
Mumbai   Nairobi   São Paulo   Shanghai   Taipei   Tokyo   Toronto

Oxford is a registered trade mark of Oxford University Press
in the UK and in certain other countries

Published in the United States
by Oxford University Press Inc., New York

© Oxford University Press, 2003

A catalogue record for this title is available from the British Library

Library of Congress Cataloging in Publication Data
Gánti, Tibor.
The principles of life/Tibor Gánti with commentary by
James Griesemer and Eörs Szathmáry.
Includes bibliographical references (p. 187).
1. Life (Biology)   2. Biochemistry   I. Title.
QH341 .G26 2003   576.8′3–dc21   2002038125
ISBN 0 19 850726 7 (alk. paper)
10 9 8 7 6 5 4 3 2 1

Typeset by Cepha Imaging Pvt Ltd

Printed in Great Britain
on acid-free paper by The Bath Press, Avon

Preface

It was sometime around the autumn of 1974 that I saw Tibor Gánti for the first time. He was delivering a series of popular lectures about the basic characteristics of life and living processes at the centre of the Hungarian Organization for the Popularization of Science. Having just started the grammar school in Budapest, already with a well-developed interest in biology, I remember how struck I was by the extreme clarity and intelligibility of those lectures. The audience was mixed, but the vast majority were students (also from the university) and schoolteachers. By then many knew that Gánti had published a seminal book in Hungarian with the title *The Principle of Life* (*Az élet princípiuma*, Gondolat, 1971). It was a very serious book in a popular science disguise. It sought to establish the foundations of the theoretical biology of individual organisms.

Trained as a chemical engineer, Gánti has always been interested in biology. In fact by the time he went to university he was convinced that the basic characterization of living systems was impossible without a firm chemical grounding. For some time Gánti worked in areas of chemical industry close to applied biology. He has filed patents for the production of biochemicals by the directed use of enzymatic biochemical networks.

However, his interest in the basics of life has never wavered. He realized that the problem could not be solved without using the new results of molecular biology. He was the first author in Hungary to popularize molecular biology (*Forradalom az élet kutatásában* (*Revolution in Life Research*), Gondolat, 1966). But already in that book there is a chapter on the basic characterization of life. Gánti argued that life at its most fundamental was a combination of two different types of process: one responsible for homeostasis (keeping the status quo in some sense) and another ensuring the controlled succession of life history events (the 'main circle'). The first relates to metabolism, the second to DNA and the genes. This 'dual nature' of life has been independently realized by other outstanding scientists (such as Freeman Dyson and John Maynard Smith). However, Gánti went far beyond that.

When he published his pioneering book on molecular biology, he was afraid that he might have to go to prison. The atmosphere of Lysenkoism was still present in Eastern Europe in the late 1960s. He had similar worries when he decided to go public with his *chef d'ouvre: The Principle of Life*. Part of the text of this volume is an updated version of that landmark book.

Why has *The Principle* been so remarkable? There are, I think, a number of good reasons. First, it built on much of the new data of molecular biology. The first edition (in 1971) contained several chapters which could be seen as a revision of Gánti's first book. Second, it approached the problems of life characteristics and units of life with unprecedented rigour for the time and in a way that is still heuristically quite valuable. Third, Gánti recognized, and *The Principle* shows, how chemical cycles (and networks) of various kinds play a crucial role in living systems and that their exact description needs a new approach (later called cycle stoichiometry). Fourth, he borrowed some key concepts from cybernetics. Fifth, in sharp contrast to the first point, he omitted a sacrosanct assumption of molecular biology. Finally, the new conceptual framework was not only presented, but was also applied to a novel attack on the problem of the origin of life.

Let me elaborate on some of these key points. Why did he adopt this Janus-faced approach to molecular biology? What is the crucial omission? Gánti thought (and still thinks) that the elementary units of life (he calls them chemotons) could be described without the incorporation of enzymes. The first argument is that enzymes just facilitate what is otherwise possible, and it is the *regulated*, rather than the *regulating*, system that is of primary importance. The second argument states that experiments related to the origin of life suggest that non-enzymatic systems precede enzymatic systems not only logically, but also historically. I suspect that a much greater number of biologists would agree with the former argument than with the latter. The question of whether non-enzymatic living systems are feasible or not is open to theoretical and experimental investigation. But the **chemoton theory** (as it is now called) by no means excludes enzymes from its framework: it incorporates them logically and historically in a cardinal way, as we shall see shortly.

Cycle stoichiometry deserves special mention. Stoichiometry is the oldest exact branch of chemistry, and so it may come as a surprise that in the order to characterize chemical cycles properly it had to be developed in a special way. The main reason is that cycles act as catalysts, and catalysts cancel out from the overall balance equations of stoichiometry. Gánti realized that the problem could not be solved without the introduction of the 'turning number', appropriately displayed in the cyclic process sign that he introduced to replace the equality sign in the familiar mass-balance equations. Although it is true that the kinetic characterization is feasible using the traditional differential equations of chemical kinetics, the equations are non-linear in such a way that only numerical solutions are possible. The cyclic process sign combines stoichiometry with a dynamical element in a shrewd way.

Also historically, stoichiometry is not unrelated to the basics of biology. Mendel's revolutionary genetic paradigm is very stoichiometric in nature: if different heritable factors are combined in fixed amounts, certain qualities are obtained. Replace 'heritable factors' with atoms, and one has the stoichiometric description of molecules.

The later development of chemoton theory saw many revised and foreign editions of *The Principle*, together with the publication of many papers and some other books, most notably *Contra Crick, or the Essence of Life* (in Hungarian, 1989), parts of which are included in the present volume. There is one crucial development which should still be considered in this Preface, namely 'priority on the RNA world hypothesis'. Biologists credit this phrase and idea to Walter Gilbert's influential *News and Views* item in *Nature* in 1986. In a Hungarian specialist journal (*Biológia*, with English abstracts) in 1979 Gánti describes the same idea (but not the phrase) in more detail and in a manner much better integrated with other problems of the origin of life. He asks how the basic non-enzymatic units of life could have acquired the capacity of enzymatically controlled metabolism. Based on preliminary ideas promulgated in 1968 (by Crick and Orgel) and in 1972 (by White), he boldly draws the picture of metabolizing reproducing chemical systems where the reactions are catalysed by enzymatic RNA molecules (now called ribozymes). The enzymatic RNAs would have been replicated as well in these ancient, but already evolved, chemotons. The genetic code and translation could have come later. In 1983 (shortly after the discovery in the United States of ribozymes acting in modern organisms) Gánti described in the same Hungarian journal how ribozymes could have been assembled in a way guided by the substrates already present. The insight related to the RNA world alone would have deserved worldwide attention.

But why did the Anglo-Saxon world miss these results? The obvious answer is that they were not published in internationally known journals. Why not? There are a number of reasons. Gánti seems to have chosen a bad publication strategy. He has always tried to promote the *grand scheme* first, and to derive specific points from that. Now, the more revolutionary an idea is, the more difficult it is to sell it, unless it focuses on one timely and specific aspect. The second reason is that Gánti has not been well funded in Hungary, to put it mildly. He has been almost completely denied the possibility of attending meetings. It is true that his English (partly as a consequence of that) is not very good, but he has been a marvellous speaker in Hungarian and an acceptable one in English. More opportunity could have helped.

The aim of this volume is to compensate for outrageous fortune as much as it is possible today. Hopefully, the thoughts of Gánti, including his grand theme, may now become better known in the Anglo-Saxon world. It is my pleasure to acknowledge and to pay back (to a minor extent) my intellectual debt to Professor Gánti by contributing to this English edition with comments and some editing. I am overjoyed to be have been complemented in this task by my dear friend and colleague James Griesemer from Davis, California. It is my conviction that 30 years after their first publication, Gánti's thoughts will still be extremely stimulating.

Eörs Szathmáry

I first met Eörs Szathmáry in the autumn of 1992. We were both new Fellows of the Wissenschaftskolleg zu Berlin (Institute for Advanced Study). I had come to Berlin expecting to turn an essay I had written,

'The informational gene and the substantial body', into a book. I had been rethinking my views on units of evolution since 1989 in light of my recent recognition that Weismann—originator of the doctrine of germ-plasm continuity and somatic discontinuity—was no Weismannian and thus that theories of evolutionary units claiming to be Weismannian, such as Richard Dawkins' replicator theory, needed reassessment. Eörs was working on his book, *The Major Transitions of Evolution*, with John Maynard Smith. Those were wonderful times, although the great achievement of my year in Berlin turned out to be a delay of 10 years in writing my own book while I sorted out these new ideas. Eörs' ideas, and his introduction to Gánti's work, led me to believe that more than fixing up Weismannism was needed to clarify my thinking. Several years later, Eörs invited me to Budapest's sister institute to the Wissenschaftskolleg zu Berlin, Collegium Budapest, to continue my quest. I owe to Eörs and to Gánti (whom I had the great fortune and pleasure to meet in Hungary) the very exciting, challenging, and frustrating struggle that I have had over the last decade to articulate a new conception of the units of evolution. At the heart of 'my' perspective is Gánti's chemoton model.

Three features of Gánti's model are of fundamental importance philosophically. First, it is a *chemical* model. The importance of introducing a seriously chemical way of thought into the philosophy of biology should not be underestimated. These days, molecular biology is prominent in the thinking of many philosophers of biology, but we know from the many and various critical reviews of molecular biology as a history of 'informational macromolecules' that this molecular biology sustains a curiously un-chemical philosophy. Gánti's work opened my eyes to the real possibility of a philosophical rapproachment of biology and chemistry far more sophisticated than the tired debate over whether biology is 'reducible' to chemical laws or not.

Second, Gánti's model is a *stoichiometric* model. The idea of cycle stoichiometry at once captures a fundamental feature of biological systems—that they are organized in cycles—and at the same time offers a way of understanding what it is to think chemically. In my own case, the chemoton model resonates because I had turned away from Dawkins' copy metaphor for replicators as an inadequate way of understanding units of heredity, development, and evolution. In its place, I put 'reproducers', entities which form genealogies by 'material overlap' of parts—what once were parts of parents become parts of offspring. While the *relations* between parents and offspring might be understood in terms of copy-similarity, the *process* of reproduction is fundamentally different from copying processes. Since Gánti's cycle stoichiometry tracks the flow of *matter* (not only of information) through chemical processes, the chemoton model seems well suited to the kind of materialist (or better, Aristotelian) basis of all our biological thinking.

Third, Gánti's chemoton model helped me break out of the dualis-tic thinking enforced by that grand triumvirate of dichotomies in modern biology: Weismann's germ/soma, Mendel's factor/character, and

Johannsen's genotype/phenotype. The simple chemoton has three constituent stoichiometrically coupled subsystems: an autocatalytic metabolism, a 'genetic' polymer, and a 'genetic' membrane. We have known since the polymath Charles Saunders Peirce that 'three-ness' is far more conceptually productive than 'two-ness'—the complexity afforded by having *three* interrelated subsystems rather than two polar opposites has helped lead me to a more nuanced view of the nature of the processes of inheritance and development by showing how these processes need *not* be treated as causally or logically autonomous (as in Weismannism's distinction of a developmental process running from genotype to phenotype and an inheritance process running from genotype in one generation to genotype in the next).

Indeed, I find myself turning frequently to the chemoton model these days as a touchstone, a heuristic model to keep my new way of thinking on course and to prevent backsliding to old ideas that cannot easily be eliminated. While it has been my pleasure to contribute to the reissue of Gánti's work in this volume, in partial repayment of intellectual debts to my two Hungarian friends, I think that the greatest pleasure will be in making Gánti's work available to a new generation of readers. I hope that they will be as astonished as I have been that a book conceived in 1968 can continue to be heuristically fruitful despite the vast changes of chemical, molecular, and biological sciences since then.

James Griesemer

Organization of the book

This book is organized into five chapters. The first is a new essay by Professor Gánti, 'Levels of life and death', which introduces the reader to some of the concepts central to his main argument as well as illustrating how the core problem of the nature of the living state applies at many levels of life.

The second chapter, 'The nature of life', is a revised version of the second part ('Solutions') of Gánti's book *Contra Crick, or The Essence of Life*, published in Hungarian in 1989. Responding to Crick's 'panspermia' hypothesis in his book *Life Itself: Its Origin and Nature* (Simon and Schuster, New York, 1981), which attributes the chemical origin of life on Earth to inoculations of organic molecules from space, Gánti offers a set of reflections on the parameters of the problems to be solved in both origins of life research and, more broadly, in the search for principles governing the living state in general.

While it may well be true that life on Earth began by inoculation from space, the problem of finding general principles of the living state and the related problem of the chemical conditions for the origins of the living state are not solved by pointing to the *place* of origin. Gánti's meditations help to reorganize the problems in terms of his chemical way of thought. This is useful in light of the fact that the informational paradigm in molecular biology heralding from Crick's work with Watson as well as his development of the 'central dogma' of molecular genetics is something to be explained, not presupposed by a theory of the origin of life or the principles by which the living state is to be recognized. Thus, in this chapter, the reader is introduced to the chemoton, to cycle stoichiometry, and to a chemical way of thought that transcends whatever chemical model of the moment may be directing (or misdirecting) our attention to fundamental principles.

The third chapter, 'The unitary theory of life', comprises the majority of the much revised 6th edition (1987) of Gánti's book *The Principle of Life* (published in Hungarian in 1971). We owe a debt of gratitude to L. Vekerdi, who translated Gánti's *Principle* into English for the first time (as well as to Erzsébet Czárán for translating *Contra Crick* and Viktor Müller for translating the Introduction into English). In this chapter, Gánti develops his theory of the chemoton in accessible language and without all the technical details of chemical kinetics. Central to this is an account of 'life criteria' which builds as the reflections in chapter 2. The life criteria serve to articulate a basic philosophy of units of life which encompasses everything that philosophers have discussed under the restricted heading of 'units of selection'. He then extends and applies the theory of these fundamental units to considerations of conventional ways of organizing discussion of life principles, such as genetics, chemical synthesis, and the requirements of what he calls 'exact theoretical biology' generally.

Chapters 4 and 5 are essays by Eörs Szathmáry and James Griesemer on the biological and philosophical significance respectively of Gánti's work. In these essays we try to indicate not only how Gánti's work has proved significant to us personally, but why we think it has continuing relevance and heuristic power for theoretical biology. Finally, we include comments by each of us on Gánti's contributions throughout the book, bringing his work into relation with current literature in biology and philosophy. Comments by Eörs Szathmáry are identified by the prefix S and those by James Griesemer by the prefix G.

Eörs Szathmáry
James Griesemer

Introduction

This book aims to unveil the secret of the nature of life, i.e. to identify the fundamental differences between living and non-living systems on the basis of the chemoton theory, and then uses the conclusions to hypothesize about the key steps of the spontaneous genesis of life. The core of chemoton theory was formulated in November 1952. Thus it is nearly 50 years old and has the perspective of half a century. If half a century provides a historical perspective in everyday life, it does so even more in general science and particularly in the science of biology, which has progressed exceptionally rapidly in the second half of the twentieth century. The theory was first published in 1971 in *The Principle of Life* (in Hungarian). Thirty years have elapsed since then. In the history of science, in the history of biology, three decades again provide a historical perspective. The validity of the basic theorems of chemoton theory can thus be judged from a historical perspective. Time is the best measure of success in science.

None of what is now known as molecular biology had been conceived in 1952. It was already suspected that heredity was somehow associated with DNA, but there was not the faintest clue to the mechanism of this association. The structure of proteins was also unclear. It was known that proteins are macromolecules composed of amino acids, but as for the structure of the molecules, the sequence of amino acids (if there was be any defined sequence at all), and how these traits determine the chemical function of the protein enzymes, biology was still in the dark.

A few metabolic pathways had already been elucidated. Above all, the chemical reactions in the biological oxidation of sugars had been essentially identified and much was known about the successive chemical transformations involving organic acids in biological oxidation. The Krebs cycle had also been described as the first biochemical cycle. Although a few autocatalytic processes had been known since the beginning of the twentieth century, these had been identified the field of chemistry and not biology. The nature of autocatalytic processes was not yet understood and no biochemical examples had been identified.

The molecular background of these processes was beginning to be elucidated in the 1950s and 1960s. The structure of DNA was determined in 1953 by the Watson–Crick model, which also revealed the molecular mechanisms for the storage and copying of information. The primary structure of proteins, i.e. the nature and specificity of the amino acid sequence, was beginning to be elucidated in the late 1950s, and X-ray studies revealed the secondary and tertiary structure, i.e. the helical structure of the amino acid chain and spatial folding, at about the same time.

These were fascinating discoveries which laid the foundations of a new discipline—molecular biology. It turned out that the storage and copying of heritable information, its transcription for enzyme synthesis, and enzymatic function depend on the molecular structure of biological macromolecules. Subsequent decades revealed how the genetic programme controls and the synthesis of enzymes regulates the functioning of living systems at the cellular level. As these discoveries coincided with the first steps of cybernetics and computer science, these two otherwise distant fields provided mutual reinforcement for each other, so that molecular biology describing the programme control and regulation of living cells became predominant in the second half of the twentieth century. The majority of biological research institutes around the world have focused on these problems and results from these fields have been favoured for publication in the scientific literature.

In the second half of the twentieth century, people have forgotten that programme control must control a functional system and enzymatic regulation must also regulate a functional system. What this functional system is has not been studied in the second half of the twentieth century. However, the secret of life is hidden in these systems—in their organization and functioning.

Chemoton theory is concerned primarily with this machinery aspect. As mentioned above, it was first formulated in November 1952. At that time it was already clear that the processes of living systems involve series of chemical transformations, and the organized regulated processes that occur inside living cells are realized by interconnected systems of chemical reactions. These chemical reaction systems must also be self-replicating (autocatalytic).

The basic idea proposed in 1952 was as follows: all living systems must contain an autocatalytic metabolic network comprised of reversible reactions and a similarly autocatalytic but unidirectional reaction system of irreversible reactions. The specific coupling of the two creates a functional system or 'machine' which, unlike machines created by humans, manipulates the energy that flows through it in chemical rather than mechanical or electrical ways. This is the basis of the functioning of all living systems without exception. This is the system that is controlled by the genetic programme and regulated by enzymatic regulation. The first chemical reaction system could already be envisaged at that time but no physical–chemical evidence had been presented for the existence of the second. A year later, in 1953, the discovery of the Watson–Crick model made it clear that the functions of the second subsystem can be performed by mechanisms related to nucleic acid synthesis.

However important, a basic idea is not a theory. To make it a theory, it has to be reconciled with observed reality and with the related fields of science, and it has to be developed into a form that can be tested by experiments and is suitable for mathematical analysis. This was not possible at the level of the science of that time and the author, then an undergraduate, could not, of course, work out the theory. It took nearly

20 years to develop a rudimentary theory from the basic idea, and this theory was published in the book *The Principle of Life* in Hungarian in 1971. In this book, the term **chemoton** was coined for the simplest chemical 'machine' which shows the generally established characteristics of life. Chemoton theory is the system of exact inferences drawn from the chemoton model. As we shall see below, the theory has been extensively, but not widely, expanded in the past decades.

As the theory unfolded, the problem of the origin of life moved more and more to the foreground. It has become clear that the description of a hypothetical minimum living system which already shows the characteristics of life but does not contain any additional mechanisms that are not relevant for the essence of life (although perhaps indispensable in a given habitat) can also be approached from the bottom up, i.e. from chemical systems, and not only from the top down. Great advances in chemoton theory took place in the 1970s. Computer simulations of the chemoton model were performed and chemical–mathematical methods were developed for the exact quantitative study of the model. It was found that a third subsystem is needed—one which encloses the whole system and ensures spatial separation, i.e. the division of the system during its duplication. This was first described mathematically without any underlying concrete physical–chemical reality. Then, in the summer of 1972, Singer and Nicholson published a paper in *Science* on the structure of two-dimensional fluid membranes which provided the physical background for the third subsystem of the chemoton model, which had previously only existed as a mathematical construct. In the second edition of *The Principle of Life*, which was published in 1978, these new results were properly incorporated and the section on the chemoton model and theory was completely revised. More than 20 years have passed since then and there have been another five editions of the book (making a total of seven), but no further changes were necessary in the text. The present book, which can be regarded as the eighth edition, contains the original text of the 1978 edition.

Since 1952, there have been an enormous number of discoveries in biology. None of them has been inconsistent with chemoton theory. On the contrary, the likelihood of several discoveries was predicted by inferences from chemoton theory and thus these have provided validation for the theory. In *The Principle of Life* the theory is approached from the notion of life. It seeks answers to the following questions. What makes something living? Why is it fundamentally different from non-living? Where is the borderline between living and non-living? What is the secret of the peculiar characteristics that are universal in all living organisms but cannot be found in any non-living system? However, as the theory developed and became more exact, it has become clear that the problem has to be more broadly generalized. It turned out that the fundamental principles behind the functioning of living systems can be understood in terms of 'fluid automata'. These are complex systems of chemical reactions which function like machines; they can be regulated and controlled, but do not necessarily contain solid parts. All the regulated and controlled processes take place in fluid phase via chemical mechanisms. Such systems are now called fluid automata (however, in *The Principle of Life*, the term 'soft automaton' is used and for historical reasons we have not changed this in the present edition). During the development of the theory, it became clear that the chemoton model, i.e. the minimum model of living systems, can also be approached from the viewpoint of chemistry. Chemical reactions as building blocks can be assembled into regulated and programme-controlled chemical automata without including any solid components.

This part of chemoton theory—the theory of fluid automata—was published by the National Centre and Library for Information and Technology (OMIKK) in 1984 as the first volume of a chemoton monograph. The second volume, subtitled *A Quantitative Theory of Living Systems*, was published by OMIKK in 1989 (both volumes in Hungarian). This extensive development of the theory has made it possible to approach the nature of living systems not only downwards, moving top down from complex biological systems towards simplicity, but also upwards, moving from chemistry towards the more complicated systems. In this way, a logical chain of events can automatically be formulated to describe the development of living from non-living systems during the geological evolution of planets. This logical deduction is presented in my book *Contra Crick, or the Essence of Life*, published in 1989. The circumstances of the writing of this book are unusual, as it was intended as a reply to Francis Crick's book, *Life Itself: Its Origin and Nature* (Futura Publications, 1982), in which he argues that biological information cannot have arisen spontaneously. Crick's argument has since become obsolete and Crick himself has abandoned the theory described in the book, so that the first part of *Contra Crick, or the Essence of Life*, which contains a criticism of his book, is no longer relevant. However, the second half, which is concerned with the origin of life, is today, more than 10 years later, even more relevant to current research. Therefore this second half is incorporated in the present volume. Indeed, it forms the first part of this book as the origin of life, described as a series of events, is perhaps more readily readable than the more philosophical *The Principle of Life*.

Instead of growing obsolete, the contents of this book have moved further into the foreground of science in recent decades. This is likely to have prompted Oxford University Press to publish this compilation in English which can be regarded as the eighth edition of *The Principle of Life*.

However, chemoton theory extends beyond the scope of this book, which only contains the logical deductions. The technical details have been published in the two-volume monograph *Chemoton Theory* (OMIKK, 1984, 1989), unfortunately up to now only in Hungarian, mentioned above. However, this is a textbook directed at those who are especially interested in the topic, and prior knowledge of chemistry, biochemistry, and biology is assumed.

I have received a lot of help in developing the theory from young scientists who joined the research in the 1970s and 1980s. Most of them were graduate students, writing their theses on the topic. Today they are successful scientists, many of them professors with an international reputation: Ferenc Békés, Ákos Nagy, Eörs Szathmáry, Máté Hídvégi, Pál Korányi, Mária Újhelyi, Tibor Csendes, László Schlemmer, László Demeter, Csaba Gáspár, and Mátyás Korpádi, to give an incomplete list, and József Verhás, who worked out the physical–chemical aspects of the division of microspheres. I am especially grateful to Professor of Physics Dr István Gyarmati, who provided me with guidance on how to make the theory exact and helped me in the development of chemoton theory. Finally, I owe a debt to Collegium Budapest for its support in the preparation of the English edition of this book, and to Professor Eörs Szathmáry and Professor James Griesemer for their help with the publication and for preparing the notes.

Tibor Gánti
Nagymaros
May 2001

Contents

1 Levels of life and death   1
   1.1   Introduction   1
   1.2   Explanation of some of the basic concepts   2
   1.3   The minimal system of life: the chemoton   3
   1.4   Life at the prokaryotic level   6
   1.5   Life at the animal level   7
   1.6   Between and beyond   9

2 The nature of life   11
   2.1   It would be useful to know what we are looking for   11
   2.2   Fluid machines   16
   2.3   Order in the nothing   19
   2.4   And who is the constructor?   23
   2.5   Chemotons   29
   2.6   Chemotons of the primordial Earth   36
   2.7   The birth of primordial texts   42
   2.8   We crossed the 'finish line'   51

3 The unitary theory of life   55
   3.1   Exact sciences   55
   3.2   The units of life   58
   3.3   Sets and systems   63
   3.4   Function and stability   68
   3.5   The criteria of life   74
     3.5.1   Real (absolute) life criteria   76
     3.5.2   Potential life criteria   78
   3.6   Subsystems of the living cell   81
   3.7   The chemical motor   85
   3.8   Growing systems   95
   3.9   Chemical systems multiplying by division   101
   3.10   The chemoton   107
   3.11   Life at the chemoton level   111
   3.12   Computer simulation of the function of chemotons   115
   3.13   Chemoton theory involves the principles of soft automata   120
   3.14   Chemoton theory involves some basic laws of genetics   124
   3.15   Chemoton theory involves an explication of the origin of life   132
   3.16   Chemoton theory involves the strategy of the synthesis of living systems   140

3.17   Chemoton theory involves the possibility of an exact
        theoretical biology   147
3.18   The responsibility of the biologist   152

4   **The biological significance of Gánti's work in 1971 and today
        (Eörs Szathmáry)**   157
4.1   *Annus mirabilis* 1971: Eigen and Gánti   157
4.2   Life criteria: units of evolution and units of life   158
4.3   Gánti's concept and some other definitions of life   160
4.4   The chemoton satisfies the life criteria   162
4.5   Compartmentalization and the problem of
        minimum life   162
4.6   The autopoietic concept   163
4.7   The chemoton: a reproducer built from replicators of
        three types   165
4.8   Metabolic replicators and the side-reaction problem   166
4.9   Enzymatic RNAs (ribozymes)   167
4.10   Protocells without a metabolic subsystem:
        the ultimate heterotrophs?   168

5   **The philosophical significance of Gánti's work
        (James R. Griesemer)**   169
5.1   Units and levels: biology and philosophy   169
5.2   Self-reproducing automata in the fluid state   175
5.3   Cycle stoichiometry: the language of living systems   178
5.4   The chemoton   179
5.5   Lessons from Gánti's work: engineering models and
        life criteria   180
        5.5.1   Control is distributed   182
        5.5.2   Stoichiometry disciplines   184
5.6   False models as means to truer theories   185

References   187

Index   195

1 Levels of life and death

Gánti, T. 2000

1.1 Introduction

Most biologists think, or at least feel, that the question 'What is life?' is only an abstract philosophical question, outside the consideration of modern biology. This point of view is not only erroneous, but is also definitely harmful for the development of the biological sciences. Of course, the nature of life *is* a philosophical question. But in addition to this, it is also a fundamental question in the natural sciences. An answer to this question is an indispensable precondition for the birth of theoretical biology. It is also a very important social question, as a multitude of social problems will remain unsolved without a concrete understanding of the nature of life.

Starting points of the natural sciences are usually some minimal model systems. These are systems which can show, during the evolutionary development of matter, the special properties particular to a given scientific discipline. For example, elementary electronic charges (protons and electrons) became the fundamental starting points for the electrical sciences, molecules for chemistry, elementary cell structures for crystallography, and genes for genetics. Until these discoveries, biology had no such minimal system. However, it is obvious that during the evolution of the organization of matter, material systems had to achieve a certain level or critical degree of complexity. Below this level systems are not alive, but above it they start to show the most general and principal phenomena of life. Gánti proposed such a minimal construction in 1971, which he called the **chemoton**. In order for the chemoton model to fulfil the role of the minimal system for biology, its organizational principles must be present in every living being and must be absent in non-living systems. If we have a model of such a minimal system, then the question 'What is life?' is no longer an abstract philosophical question. It is a basic question of the natural sciences and the answer has a scientific basis; it can be described with exact mathematical methods and its correctness can be verified by concrete experiments. Such a model makes it possible to obtain exact answers for the question of the origin of life, and in the absence of such a model we do not even know what we are searching for.

In addition, human societies have a multitude of problems which cannot be solved without exact knowledge of the nature of life. Some of these are really urgent today: genetic manipulations (especially on human embryos and embryonic tissues), transplantation, abortion, and the question of clinical death. These questions have many legal, ethical, moral, and religious consequences which cannot be addressed without understanding what is living and what is not.

1.2 Explanation of some basic concepts

'What is the characteristic feature of life? When is a piece of matter alive?' asked Erwin Schrödinger in his famous booklet *What Is Life?* His answer was: 'It is alive if it goes on "doing something", moving, exchanging material with its environment, and so forth, and if it is doing that for a much longer period than we would expect an inanimate piece of matter to "keep going" under similar circumstances'. This description really characterizes one of the most fundamental features of living things. All living systems—while alive—do something, work, function. A system that does not do anything is not alive. It may be lifeless, which means that it does not live, never was alive, and is not ever going to live. It may be dead, which means that it may have lived earlier, but has lost this capacity. Or it may be in an inactive state, which means that it would be able to live but this function does not work at the given moment (e.g. frozen bacteria or dried seeds). It may start to live again under suitable conditions. Schrödinger's characterization really referred to the most important property of living beings.

However, living things are not the only systems that 'do something' and also do it for long periods. For instance, the wind may blow for prolonged periods and rivers do their erosive work continuously. In technology, engines, vehicles, and automata are also able to work without interruption. What is common to these systems is that they are positioned between the higher and the lower potential level of some kind of energy, and, part of the energy which flows through the system is transformed to work.

However, there is an essential difference between the work performed by the wind or a river and that performed by some machinery constructed by humans or by living systems. The work of natural forces is undirected: it is caused by some temporary environmental condition and produces in random changes in the environment. The work of machines and living systems is directed; it is programmed by some inner structure, i.e. by their construction, and produces useful work for the creator (humans) or for the living entity.

To get from random to directed work, the flow of energy must be manipulated along a series of forced trajectories within the system. Mechanical machines are manipulated by mechanical means (e.g. wheels, axles, springs, pendulums), and electronic apparatus is manipulated by wires, coils, rectifiers, capacitors, etc. The same procedure occurs if the driving force of the machine is chemical energy, as is the case with steam engines or battery-driven electric apparatus. Here, the chemical energy is first transformed into mechanical or electric energy and this is subsequently manipulated in forced trajectories by mechanical or electronic means.

The driving force of living systems is chemical energy. This is also true for plants, in which the energy of the light is initially converted into chemical energy. However, in contrast with the situation for mechanical machines, the energy flow in living systems is manipulated by chemical means. This is the most important difference between mechanically constructed machines and living systems. It is also the source of many special properties of living systems, such as, the capacity for spontaneous formation, growth, and regeneration, but above all the capacity for reproduction. In contrast with manmade technologies, where the machines are based on mechanical or electronic automata, living systems are fundamentally chemical automata. During evolution, the mechanisms of living systems have sometimes been extended using mechanical and electronic components, but their basic structures remain chemical automata. They manipulate the driving energy by chemical methods.

As chemical reactions can proceed with suitable intensity only in the fluid phase (gas or solution), a chemical automaton can only work in the fluid phase, i.e. the continuous presence of some kind of solvent is essential. This is why chemical automata are also known as 'fluid automata'. The functioning of mechanical automata is restricted to a rigorous geometrical order of their parts, and the functioning of electronic automata is also restricted to some geometric arrangement of their components. The functioning of the fluid automata is largely independent of any kind of geometrical order. It works even if the solution is stirred, or if half of it is poured into another container, etc. Compared with mechanical and electrical automata, i.e. the 'hard automata', these properties provide living systems with highly favourable possibilities. One of these is, the capacity for reproduction—autocatalytic systems are well known in chemistry. In the following sections, we shall consider living systems as fluid machines. Higher levels of evolution produced living systems with hard and partially hard (so-called 'soft') components for the manipulation of the flow of mechanical and electrical energy. Nevertheless, because the fundamental functioning of these systems depends on chemical manipulation, we shall refer to them as soft automata.

1.3   The minimal system of life: the chemoton

There are some examples of the existence and use of fluid automata in human technology, but these are mostly known as chemical reactions (or reaction systems) and not as fluid automata. Examples of such fluid automata include the well-known families of oscillating chemical reactions or the biochemical systems used in chemical industry to produce sugar phosphates. But the fundamental unit (i.e. the minimal system) of biology must have some specific properties:

- it must function under the direction of a program
- it must reproduce itself
- it and its progeny must be separate from the environment.

Gánti first described such a model in 1971 and he continued to develop the idea in further publications.[G1] He named these fluid automata **chemotons**.

[G1]Other authors working on concepts of minimal systems include Morowitz *et al.* (1988).

[G2]In my view, reproduction is the propagation of developmental capacities by transfer of material parts of parents to offspring (Griesemer 2000a, b, c) The involvement of development in this definition of reproduction means that reproduction and autocatalysis can only be synonyms if the latter also involves 'development'—chemical development.

A chemoton consists of three different autocatalytic (i.e. reproductive) fluid automata, which are connected with each other stoichiometrically.[G2] The first is the metabolic subsystem, which is a reaction network (optionally complicated) of chemical compounds with mostly low molecular weight. This must be able to produce not only all the compounds needed to reproduce itself, but also the compounds needed to reproduce the other two subsystems. The second subsystem is a two-dimensional fluid membrane, which has the capacity for autocatalytic growth using the compounds produced by the first subsystem. The third subsystem is a reaction system which is able to produce macromolecules by template polycondensation using the compounds synthesized by the metabolic subsystem. The byproducts of the polycondensation are also needed for the formation of the components of the membrane. In this way, the third subsystem is able to control the working of the other two solely by stoichiometrical coupling. As they work, the three fluid automata become a unified chemical supersystem through the forced stoichiometrical connections. This means that they are unable to function without each other, but the supersystem formed by their co-operation can function. It can even proliferate in a rigorous geometrical sense, as has been shown by Gánti in *The Chemoton Theory*.

The chemical construction of the chemotons is illustrated by the chemoton model (Fig. 1.1), which is the most elementary simplified

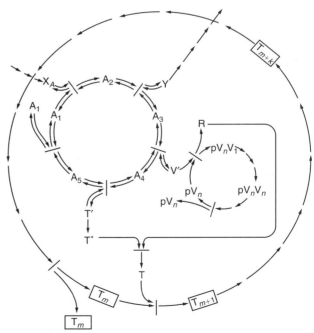

Fig. 1.1 Minimum model of chemotons. Three self-producing systems are coupled together stoichiometrically: cycle $A \rightarrow 2A$, template polycondensation $pV_n \rightarrow 2pV_n$, and membrane formation $T_m \rightarrow 2T_m$. This coupling results in a proliferating program-controlled fluid automaton, known as the chemoton.

description of chemotons. Presumably it cannot be realized in this form, but this simple model demonstrates all the indispensable stoichiometrical connections which must be present in every kind of chemoton construction. Kinetic analysis of the elementary chemical reactions allows us to perform an exact numerical investigation of the workings of the chemotons using a computer. These investigations have shown that chemotons behave as living beings, i.e. they satisfy all the criteria which characterize living systems in general (life criteria).

The chemoton model is an abstract model. By using it we can understand how it is possible to organize a chemical supersystem from several autocatalytic subsystems, which are directed by a central program, and which can reproduce itself. The fact that it is an abstract system means that its components are not restricted to particular chemical compounds. However, they must have certain stoichiometric capabilities and, they must be able to produce certain compounds, which are important for the whole system. Any of the simplified subsystems of the model can be replaced by a complex real chemical network if it is able to ensure the required stoichiometric connections. Our working group has constructed such a realistic system, consisting of more than 100 different organic components. Some of these components are known to be materials which existed prebiotically; the existence of others under such conditions is possible. This realistic model (at least stoichiometrically) could work as a program-controlled reproductive chemical supersystem, using well-known chemical compounds, which existed prebiotically, as nutrients. Máté Hidvégi constructed the major part of the chemical network and Eörs Szathmáry performed the stoichiometrical calculations. Of course, this realistic prebiotic model is not the only possible one; there are many other possible solutions. Because there are no chemical restrictions on the realization of the model, it could even consist of silicon compounds. The only prerequisite is that the silicon chemistry should be able to produce reaction systems suitable for the stoichiometric criteria of the chemoton model. Therefore the chemoton model is able to represent any type of living system, even extraterrestrial ones.

The chemoton model is designed from the bottom up, starting from the direction of the chemical reactions. Linkage of three autocatalytic subsystems has resulted in a supersystem which is controlled by a program and is able to reproduce itself. Appropriate investigations have shown that this construction is able to satisfy the criteria for life. It is probable that chemotons are the simplest systems possible which show the properties of life. Consequently, the chemoton model, as the simplest system capable of satisfying the criteria for life, can be regarded as the minimal system of biology.

The chemoton model has a further significant property. In the model only stoichiometrical connections are defined, i.e. only the quality and quantity of the chemical components and the pathways of their formation, transformation and decomposition are described. The model does not contain any prescription or restriction on the speed of the chemical reactions in the system. Therefore it remains valid whether the reaction rates are determined exclusively by the concentrations of the components or are influenced by catalytic effects, or even if the processes are accelerated by complicated enzymatic systems, as happens in the living world today.

1.4   Life at the prokaryotic level

If we are searching for the secret of life—trying to define the most fundamental differences between the animate and the inanimate—then we have to investigate the simplest and most primitive level of life existing in the world today. According to our present knowledge, this is represented by the prokaryotes. The world of the prokaryotes is now extremely diversified, as they have developed to their present level of diversity over a period of billions of years. Even so, they have common properties which are the fundamental characteristics of their being.

The simplest known prokaryotic organisms are the thermoplasmas and the mycoplasmas, which are completely fluid organisms bounded by two-dimensional fluid membranes. There are two subsystems inside the boundary membrane—the cytoplasm and the genetic material. The cytoplasm is essentially the metabolic subsystem, consisting of small-molecule metabolic networks which produce all the chemical components needed for the working and reproduction of the total system. These components include the building blocks for membrane formation as well as for the reproduction of the genetic materials. The genetic material contains the genetic information for the whole system. Therefore the structural bases of the myco- and thermoplasmas correspond very well to the chemoton model. They have the three subsystems required—the metabolic, the membranous, and the informational. These three sub-systems have no solid components; everything, even the boundary (membrane) subsystem, is in the fluid state. However, they may contain additional subsystems which have been developed during evolution. Such subsystems include, for example, the protein-synthesizing subsystem or, in some cases, the flagella which are semi-solid structures.

The world of the prokaryotes is very diverse. They include many different metabolic types, ranging from the sulphur bacteria to the methane-producing bacteria, from the obligate anaerobes to the aerobes, and from halophilic to hyperthermophilic organisms. They also have an enormous variety of morphological properties. They can be formless, spherical, stick- or thread-like, or spiral. They can be motionless or fast-moving. Some are surrounded by fluid membranes only; others have solid cell walls or even a slimy shell enclosing their cell walls. But the fundamental construction of the three subsystems is present in every one of them. During evolution several additional parts and subsystems have been added to the fundamental construction without disturbing the co-ordinated functioning of the three original subsystems.

The chemoton model can be considered as a suitable representation of living prokaryotic systems. We can state that, at the level of the prokaryotes, only those systems which are based on the fundamental properties of stoichiometric organization, defined by the chemoton model, are alive. Although many different types of metabolism are known, which is what makes it possible for prokaryotes to live in a variety of extreme environmental conditions, each of them must have an autocatalytic metabolic network to produce the raw materials for membrane formation and to reproduce the information subsystem, i.e. the genetic subsystem. It is also true that both the boundary membrane and the information subsystem (genetic apparatus) are autocatalytic and self-reproducing, and that their co-operation results in a chemical supersystem, where the co-ordinated functioning comes primarily from the stoichiometric connections between the three different subsystems. Everything else, including the chemical kinetic relations, is based on these stoichiometric connections.

The existence and co-operation of these three subsystems is the prior condition for the presence of life at the prokaryotic level. If any of them is absent, the system is no longer alive. In the case of viruses, only the information subsystem and, for more complex viruses, the boundary membrane are present—the metabolic subsystem is missing. Viruses cannot function by themselves—they cannot grow and they cannot multiply. They can reproduce only by entering cells and forcing them to make copies to produce identical viruses.

Similarly, one may make lysates from prokaryotes, i.e. material in which the cell membranes have been destroyed. These lysates can be used to synthesize biochemicals, because the metabolic subsystem still works for a time. However, these are no longer living systems; they do not have the organization of the chemotons and they cannot grow or proliferate. Their functions are not directed by an information subsystem, but only by the general rules of chemistry.

Thus, with the help of the chemoton model we can define clearly the difference between living and non-living systems at the prokaryotic level.

1.5 Life at the animal level

At least two different kinds of life exists. I would like to repeat and emphasize this: we can identify at least two different kinds of life. I must emphasize that this statement is not a hypothesis but the everyday experience of biology and medicine. However, to the best of my knowledge, nobody has yet recognized, or at least stated, this fact. Because this important fact has not been clearly stated, the concept of the two kinds of life is confused with other phenomena in both scientific and everyday applications. We know that after the death of an animal its organs, tissues, and cells remain alive for a time. This is why human organs can be transplanted from a death to a living person, and why it is possible to produce tissue and cell cultures. If we consider death as nothing other than the irreversible end of a life, then logically we must conclude that the life of a human or an animal and the life of its organs, tissues or cells are not the same.

There is also a difference in construction the two kinds of life, i.e. the life of an animal and that of its cells. Life at the prokaryotic level is characteristic of systems which are organized directly from chemical processes into living entities. Life at the metazoan level is characteristic of systems which have been organized into living entities directly from cells, which are also living entities. These systems, where the matter has a higher hierarchical organization than in the cells, use chemical as well as mechanical and electronic processes for energy utilization.

Life itself is always the property of an entity. This means that, for example, it is impossible to cut a prokaryotic cell or an animal in half in such a way that both halves remains alive. There are some exceptional cases (e.g. annelids) which can be explained.[S1]

Perhaps we could define life at the prokaryotic level, which we have characterized using the chemoton model, as **primary life**. Life at the metazoan level, i.e. the life of those 'biological supersystems' whose elementary units are living cells themselves, can be defined as **secondary life**. It is significant that the three-subsystem structure defined by the chemoton model is also characteristic of systems having secondary life. Each animal entity has a subsystem governing the geometrical structure (skin, bones), each has a subsystem governing the metabolic processes, i.e. the supply and transformation of nutrients for the whole animal (digestive tract, secretory organs, blood circulation), and each has an information and control subsystem (the nervous system) to ensure the co-ordinated working of the organs under given internal and environmental conditions. Naturally, as with prokaryotes, several additional parts or subsystems may be added to the three fundamental subsystems to satisfy special requirements or living conditions. It seems that the three-subsystem organization of the chemoton model, which leads to the phenomena designated as life at the prokaryotic level, also leads to similar phenomena at higher levels of the organizational hierarchy of matter. Such phenomena have been designated as 'life' in several languages during historical times.

Each living animal represents two kinds of life at each moment of its existence: the primary life of its cells and the secondary life of its body. The secondary life is the real life of the animal. However, the ovum, which contains all the information about a living animal,[G3] does not show or contain any phenomena related to its secondary life. Therefore, we must conclude that the life of an animal does not start in its ovum or at fertilization, but begins during embryonic development as the information subsystem (the nervous system) begins to control the functioning of the two other subsystems. In the same way, the animal dies at the moment that its information subsystem irreversibly ceases to control the other two subsystems. This is why, in medical practice, assigning the moment of death to the irreversible cessation of brain function is correct. And it is also why considering the fertilized ovum as a living human being is incorrect. Human life only starts at the moment when the central

[S1]Since Gánti does not provide an explanation, it seems worthwhile to draw attention to the recent development of the concept of individuality. Santelices (1999) gives an excellent survey of individuality concepts. Three criteria have repeatedly been used in the literature: (i) genetic homogeneity, (ii) genetic uniqueness, and (iii) autonomy. Santelices suggests that, by combining the presence and absence of these traits, eight possible kinds of individual are possible. For example, cells in animals or workers in insect colonies have traits (i) and (ii) but lack trait (iii). It is remarkable that some clonal red algae and fungi lack all traits, since they are chimaeric and modular at the same time. This clearly reveals that some additional feature must qualify them as organisms, since it does not make sense to say that they are organisms because they are not genetically unique, not homogeneous, and they lack autonomy! Gánti's observation about the presence or absence of a controlling subsystem adds a new dimension to Santelices' system. However, it must be recognized that the nervous system may or may not be cut into two halves: the question is really whether it is a hierarchical, centralized, or a web-like network. Web-like networks are usually divisible!

[G3]This claim is disputed in the 'developmental systems' literature; see Oyama (2000), Griffiths and Gray (1994), and the essays in Oyama *et al.* (2001).

nervous system begins to control the complete collection of embryonic cells during the development of the embryo. Laboratories working on artificial insemination discard fertilized but unused eggs from time to time, but this practice cannot be regarded as a massacre.

1.6 Between and beyond

The prokaryotes and the metazoans are only two parts of the living world. A considerable part of the living world lies between them: the enormous variety of eukaryotic cells and fungi, and the whole world of plants. So far, we have not investigated these parts of the living world from our point of view. Nevertheless, we can draw some general conclusions on the basis of our present knowledge.

The organization of eukaryotic cells is in accordance with the chemoton model. It is not a contradiction that their information subsystem (genetic material) is in a special unit (nucleus) covered with a membrane; the chemoton model defines only the stoichiometric connections and says nothing about the geometrical structures. The stoichiometric requirements are satisfied in eukaryotic cells in exactly the same way as they are satisfied in the prokaryotes. The presence of the cell organelles with their own information subsystems (chloroplasts, mitochondria) could cause us some problems, but their presence is well explained by endosymbiotic theory. Eukaryotic cells do not appear to have any type of secondary life which can be terminated without the termination (death) of the primary life of the cell.

In the multicellular organisms, an interesting situation exists in the case of the fungi and of the plants. They have two subsystems, one governing the geometrical structure, and the other governing the metabolic processes. The connection between the two subsystems corresponds to the requirements of the chemoton model. In these subsystems, just as in the case of the animals, the building components are not direct chemical reactions but cells with their own (primary) life. We would expect these organisms to have some sort of secondary life. However, experience shows that such a secondary life does not exist in plants and fungi: it is impossible to kill them in the way that animals can be killed. Animals cannot be cut in two so that that both parts remain alive; plants can be cut into many pieces, each of which is still living. Some plants are propagated in this way in agriculture (vegetative proliferation). This is because no secondary information subsystem has evolved in plants and fungi. All information about plants and fungi is stored in the DNA of their cells and no new information subsystem has developed above all level. Therefore the life of plants and fungi is no more than the co-ordinated collaboration of the primary life of their cells. Human experience mirrors this fact: several languages have words to name the act of killing people or animals, but none for killing plants or fungi. In the same way, some oriental religions care for the life of animals but not for the life of the plants.

What is at the level above the organization of animal life? In the same way as eukaryotic cells, as living units, are the building elements of a system, developing a higher level of the organizational hierarchy of matter, it is possible that co-operative functional organization can develop between multicellular organisms. This is again novel; even higher levels of evolutionary organization may be created using multicellular organisms as building elements.

Evolution has already started to proceed in this way. As a first experiment it produced families, and subsequently troops, herds and hordes evolved from several groups of animals. Perhaps the insect societies represent the highest level in this evolutionary process. For example, an ant-hill shows such a high level of organization that even its own metabolic processes appear in it: ants bring food into the ant-hill, raise fungi in it, breed plant lice, remove the waste materials, etc. All these activities equate to a metabolic subsystem on this level of the organizational hierarchy. Geometrical separation is also present, as the ant-hill is their well-defined territory. We can see that two of the three subsystems of the chemoton model are already present in the ant-hill as an entity. The question is whether they have any information or control subsystems. Several languages speak about the 'life' of the ant-hills, but this is not proof of the existence of some form of tertiary life above the primary and secondary levels.[S2]

With the appearance of humans, supra-individual systems, which clearly possess the organization characterized by the three subsystems of the chemoton model, have evolved. These are the nation states, developed during human history. Their metabolic processes are industry, agriculture, traffic, trade, etc., they have well-defined borders, and they also have an informational-controlling subsystem in the form of jurisdiction and the executive authorities. Could it be possible that the formation of nation states during history is nothing but the most recent experiment of evolution to create systems with tertiary life, in addition to primary and secondary lives? Who knows?

[S2]It is partially justified to speak of 'superorganisms' in this case. They are a special kind of individual, with segregation of 'soma' and 'germ line' (workers and queen respectively).

2 The nature of life

Gánti, T. 1989b

2.1 It would be useful to know what we are looking for

This question arises because—however unbelievable this is—science looks for the genesis of life in such a way and a manner that, in fact, it does not know what is it looking for. Namely, biologists do not wish to answer the question of what life really is. This reluctance originates partly from ignorance and partly from incredulity. Whatever miraculous details discovered concerning the mechanism taking place in living beings, biologists have the feeling that the secret of life cannot be solved. Or, if not, then they identify it with the most up to-date discovery of the given period: in the late eighteenth and early nineteenth centuries with 'organic' compounds, at the end of the nineteenth century with protoplasm, at the beginning of the twentieth century with proteins, in the middle of the twentieth century with cellular structures, and at present with DNA and the genetic information enclosed in it. These beliefs, sometimes proclaimed openly and sometimes only as guesses, have fundamentally influenced the course of research.[S3] And we should say—in the wrong direction.

In the past three decades more than a hundred symposia, conferences, and congresses dealing with scientific explanations of the genesis of life have been held worldwide, and at least as many books on this subject have been published.[G4] The expression 'genesis of life' occurs in the titles of thousands of publications, and for almost two decades a journal has appeared with the title *Origins of Life*. And what are they dealing with? The majority of them discuss the spontaneous formation of organic compounds (amino acids, dicarboxylic acids, sugars, nucleotide bases, etc.), and the rest with the formation of macromolecules (proteins, nucleic acids) and membranes (microspheres, marigranules, etc.), or even with the origin of genetic information or the genetic code. However, they do not deal with just the origin of life, other than in an introductory text according to protocol, in a superficial manner, in some philosophical disguise.

A few experiments aiming consciously or unconsciously, admittedly or secretly, at the discovery of the origin of life have been reported, although sometimes without achieving any result. Examples include Fox's microspheres and the prebiotic formations of Folsome or the Jeewanus of Bahadur. However, if a researcher finds that the formations obtained in his/her experiment are living, and they even dare to say this publicly, the answer is usually a sarcastic smile or a contemptuous pat on the back, and the researcher immediately excludes him/herself from the scientific world as a silent lunatic. As we shall see, it is not at all impossible that spontaneous biogenesis has taken place in some of these experiments, i.e. there was a *de novo* genesis of life, or at least the major part of the process leading to it. Nobody knows what should be measured on

[S3]Currently, some maintain the view that a self-replicating RNA would be essentially a 'living molecule' (Joyce 1994). See Chapter 4 in this book for details.

[G4]Reviewed by de Duve (1991). Some important original research papers are reprinted in Deamer and Fleischaker (1994).

a spontaneously formed sphere with a diameter of a thousandth of a millimetre in order to decide whether it is alive or not. Nobody knows what should be measured, as in fact nobody really knows what life is.

Biology is the science of life. There is another science, physiology, which is also considered to be the science of life. Nevertheless, there is not a single biologist or physiologist dealing with the fundamental nature of life, or at least not really seriously, comprehensively, and in an appropriate depth and scientific exactness. In the history of science there are only two scientists who really sought the essence of life, but neither of them is famous for this. One is Gottfried Wilhelm Leibniz, the scientist and philosopher who discovered the differential calculus in the late sixteenth and early seventeenth centuries, and the other is Erwin Schrödinger, the Austrian atomic physicist and Nobel Prize winner who founded the wave theory of the microworld. Of course, both these scientists explained their views about life in the language of their time and profession; this is why biologists often fail to recognize the implications of their message.

Today's scientists, depending on their profession, express their thoughts in the specific language of chemistry, cybernetics, quantum mechanics, molecular biology, etc., without understanding each other. However, it will be quite understandable to many if we say that the living system is a program-controlled cybernetic system. This is not only understandable, but also acceptable, since cybernetics itself originated from the study of the regulated and controlled operation of living systems, and program control is already familiar from the genetic program. However, if Leibniz had noticed this, how could he have expressed his thoughts, since words such as program control, feedback, cybernetics, etc. did not exist at that time?

Leibniz described living beings as automata; he called them natural or divine automata.[G5] Modern scientists smile disparagingly when hearing this; they call this mechanical materialism and use the adjective 'primitive' in relation to it. They claim that mechanical materialists thought that all kinds of wheels and transmissions were operating in living systems.

Certainly, some of the followers of these philosophers read their texts literally and were unable to grasp their meaning. However, the great thinkers did not mean that living systems were operated mechanically—at least Leibniz certainly did not. Their only means of demonstrating the assumption that living systems are cybernetic systems with controlled operation was to compare living systems with known similar systems, i.e. machines and mechanical automata. At that time, the only known continuous operations of a controlled and regulated nature occurred in man-made machines or automata. Thus living systems had to be compared with mechanical systems.

However, Leibniz recognized that there was a significant difference between man-made automata and living systems. He writes about this in section 64 of *Monadology* as follows:

... all the organic bodies of a living being are nothing more than a kind of a divine machine or natural automaton, which exceeds infinitely all artificial automata.

[G5] G.W. Leibniz, *The monadology* (trans. R. Latta), section 64 (http://www.knuten.liu.se/~bjoch509/works/leibniz/monadology.txt).

Since a machine constructed by human art is not a machine in all of its details, and e.g. the teeth of a brass wheel have details and parts which do not show any artifice at all, and they have nothing which would disclose the function of the wheel. However, the machines of nature, i.e. living bodies are machines up to their smallest details, up to infinity. This is the difference between nature and art, or more precisely, between the art of God and ours.

Of course, it should be remembered that the limit of 'infinity' at the time of Leibniz was about a hundredth or a thousandth of a millimetre (in the direction of infinitely small) and that the scientists of this time knew nothing about the cellular construction of animals, let alone the mechanism of inheritance and life processes. At present, the most up-to-date biologist also talks about the mechanism of inheritance, of translation, of protein synthesis, of membrane transport, etc. in this 'infinitely small' world. Nevertheless, today's molecular biologists cannot be called 'mechanical materialists' in the pejorative sense since, in the world of macromolecules, molecular events proceed via mechanical events in the strict sense of the word 'mechanical', except that the size of the operating parts is of the order one ten-thousandth or one hundred-thousandth of a millimetre. In what follows, we shall consider how the operation of these tiny parts is combined into a system with a controlled and regulated operation, i.e. into an automaton or, more precisely, a fluid automaton.

It was as easy for Leibniz to free himself from the idea of the 'wheeled' automaton, as it was difficult for him to cope with the control mechanism. He observed that the living system, as a natural automaton, not only operates, but also operates in a directed way. Moreover, this control does not come from outside but from inside the automaton. Thus in a living system there must be an operating part (automaton) as well as a controlling part. But what is this controlling part? What is its nature and how does it function?

These questions could not be answered at the time of Leibniz, but conclusions could be drawn about the existence and functioning of such a unit. Now we know that in animals this control is provided by the nervous system (but we do not yet know the full and precise nature of brain functions), and at the cell level it is provided by the genetic material DNA through the information encoded in it. This information is carried by the base sequence of DNA.[G6]

Of course, the sequence of signs is not material. But neither is it independent of material, since the sign is carried by some material substance (in this case by the molecular structure of a purine or pyrimidine base). Is it then material or not?

Even now this question is not easy to answer in any philosophical depth.[S4,G7] It can be imagined what an enormous problem it would have been at the time of Leibniz, when not only was the control mechanism unknown, but also the fact that a living being is a controlled system. Scientists only suspected or concluded from its behaviour that it might function this way. But the dualism in living systems between controlled and controlling parts was sensed, and perhaps the concepts of body and soul in religions also have their roots in this idea.

[G6]Modern investigations of epigenetic inheritance have shown that it is unlikely that all 'genetic' information is carried by the base sequence of DNA, regardless of whether epigenetic controls have evolved to regulate development, to control parasitic DNA, or for some other reason (Jablonka and Lamb 1995; Yoder *et al.* 1997). It is not clear to what extent the argument in the text is compromised by the existence of epigenetic inheritance systems.

[S4]Remarkably, some philosophers, such as Mahner and Bunge (1997), would like to expel the concept of 'information' from biophilosophy. This is a heroic misunderstanding at best (Szathmáry 1998). In principle, the semi-conservative replication of DNA would allow the same DNA strand to be passed on to offspring many generations ahead. It just needs a reasonable amount of luck to avoid mutations and the stochastic loss of the old strand in each generation. Yet this is a misleading picture. DNA is under continuous attack by the adverse effects of the chemical milieu, and it is rather inconvenient that water and oxygen belong to detrimental set of molecules. Therefore, if an organism is to maintain its genetic information, DNA damage must be constantly surveyed and repaired (Alberts *et al.* 1994). Even in our extreme example it is not the same *molecular* structure, but the same *information* that is conserved. Even if it may annoy some distinguished biologists, Plato was exactly right in observing that in everything alive form is much more persistent than matter. *Form* in biology arises

(contd.)

as result of the interplay between physicochemical forces and genetic information. It is amusing to compare this observation with a remark that Mahner and Bunge make against Williams, who claims that 'A gene is not a DNA molecule; it is the transcribable information coded by the molecule' (Williams 1992, p. 11)—'This is good old Platonism in modern informationist garb' (Mahner and Bunge 1997, p. 339). Whereas Williams' statement is somewhat extreme, Plato is much closer to reality in this case than Mahner and Bunge, writing 2500 years later (Szathmáry 2001a).

[G7]See Maynard Smith (2000) and Griesemer (2000c).

[G8]G.W. Leibniz, *The monadology* (trans. R. Latta), section 14 (http://www.knuten.liu.se/~bjoch509/works/leibniz/monadology.txt).

[G9]Muller (1922) has explored the fundamental physical character of the gene.

[G10]Timoféef-Ressovsky *et al.* (1935). For historical surveys of the role of physics in molecular biology, see Olby (1974, Chapter 15) and Kay (2000).

Leibniz also, struggled with the description and explanation of the 'other' part, which he sometimes called an acting principle, *entelechia*, and at other times a soul. But he drew a strict line between this 'soul' and the bodiless immaterial soul of spiritual character:

...organic bodies are never without soul and souls are never fully separated from some organic body... Thus I do not allow for naturally separated souls, neither for created spirits free of any body, and in this respect I agree with the idea of some old Fathers of the Church".[G8]

It would be very interesting to analyse in detail what Leibniz wrote about this soul, this *entelechia*—how he tried to analyse its properties, how he attempted to explain the difference between souls only 'feeling' and those already with 'intelligence', etc.—but this is not possible here as we are not dealing with the history of science in this book. Instead, we jump forward 250 years and consider how another great thinker, in the middle of twentieth century, saw life with his modern exact way of thinking.

This later thinker, Erwin Schrödinger, wrote his booklet entitled *What is Life?* in 1944. At that time 'infinitely small' had become a billionth of a millimetre or even less, down to the world of elementary particles. The laws of inheritance were known as well as their sudden variation—the mutation. It was understood that mutations originated from changes in the material of chromosomes, but nothing was known about the nature of hereditary information. Schrödinger was the first to highlight the fundamental character of these variations.[G9]

His method was agreeably simple, correct, and exact. Based on the work of Delbrück, Timofeev, and Zimmer, he calculated from the frequency of mutations caused by X-irradiation that the X-ray photon has to hit a tiny target within a volume of a fifty-millionth of a cubic centimetre in order to produce a hereditary change.[G10] This volume is of the order of magnitude of a molecule. Thus hereditary information has to be carried by groupings of atoms within molecules. But how?

Let us give the answer in Schrödinger's words.

A well-arranged group of atoms, which is resistant enough to keep its arrangement is the only imaginable structure providing a possibility for a multitude of arrangements (isomers), and which is large enough to unite the complicated system of 'determinations' in a small volume. The number of atoms in the structure should not be very large in order to ensure an almost infinite number of arrangements. To illustrate this, let us think of the Morse code. By using only two signs, the point and the dash, in arranged groups containing not more than four signs, the number of different possibilities is 30. If we introduced a third sign, and used groups not larger than 10 members, the number of different 'letters' would be 88 572.

Schrödinger called such molecular structures aperiodic crystals, and they are often referred to in the literature, although to an ever-decreasing extent.

This was written 9 years before the discovery of the molecular structure of DNA. The Watson–Crick model justified this assumption: the 'aperiodic crystal' of inheritance, DNA, contains four signs which are combined into groups (genes) of several hundred signs. It is understandable that the variety of possibilities is effectively inexhaustable and Schrödinger's conclusion is true: '...it is not unimaginable that the miniature code would describe a very complicated, given plan of development, and that it contains also tools for making it operate'.[G11]

Thus the solution which Leibniz was desperately seeking, or at least its fundamentals—the principle of program storage—was found by Schrödinger. However, Schrödinger was much too deep a thinker to have believed that the program, and especially the 'aperiodic crystal' enclosing it, would be life itself, as some of his later followers believed and even state about the concretized aperiodic crystal DNA.[G12] Schrödinger knew that, however important and indispensable the program might be from the viewpoint of life, life itself was quite a different thing. It is best to cite his own words again:

What is the characteristic feature of life? When do we say about a piece of matter that it lives? If it 'does something', moves, stands in a metabolic connection with its environment, etc., and when it 'does' this for a longer time than would be expected from lifeless matter under similar conditions.

Thus the idea of a machine, an operating controlled system, also appeared in Schrödinger's thinking. We have already mentioned that the concept of aperiodic crystals is cited in the literature. The concept of negentropy is even more frequently cited as the means by which Schrödinger characterized the operation of living 'machines'—living systems—thermodynamically. However, his comment stated above that living is only something that also does something—that operates—and that the living system is an operating system, i.e. that life is a specific operation, is never mentioned.

The essential point is that life is a specifically operating system which also operates in a program-controlled manner. Consequently, a living system should necessarily comprise at least two systems, of which one is the controlling unit and the other is the controlled part. Neither system is living if the other system is absent. It is naïve to believe that the genesis of life can be clarified by studying whether a program can be developed by itself. It cannot. Controller and controlled belong to each other, like a tape recorder and the tape. Not only can they not operate without each other, they cannot be developed without each other either. Or can they? The first tape recorders functioned not with tapes, but with steel wires. Tape recorders could function in this manner, although in a very primitive way. However, tapes would never have been developed if tape recorders had not existed.

When Crick (1981) and others ask what the probability is that DNA carrying genetic information can form spontaneously in the Universe, it is similar to asking what the probability is of developing a cassette containing an entertaining program without the discovery of the tape recorder (or any other similar device using tapes). The only answer is 'none' as the assumption itself is not reasonable. Similarly, the origin of

[G11]The variety is effectively inexhaustible because the number of *possible* combinations of nucleotides in a sequence the length of a typical gene vastly exceeds the number of molecules in any *actual* population, i.e. heredity in such a system is effectively 'unlimited' (Jablonka and Szathmáry 1995).

[G12]For example, James Watson in his famous remark that if we had the human DNA sequence we would know 'what it is to be human' (Watson 1992).

the genetic code cannot be studied without studying the origin of the reading machine using this code. The two can only function together; similarly to a living system, they originate and develop together. In complete contrast to the sequence in which science studies it today, it was not the program which appeared first, but the machine. This is because a machine can exist without program control, and even without any program at all, but the converse is not true. Later, the development of machines could promote the development of program control. The evolution of human technology took place in the same way. Therefore, let us study the process of the genesis of life from the other side—from the side of machines.

2.2   Fluid machines

When analysing the thoughts of Schrödinger we stated that the essence of life is in a specifically operating system, which is also program-controlled. The basic principle of molecular storage of the program was discovered by Schrödinger, and its concrete mechanism was provided by Watson and Crick when they clarified the structure of DNA. But what is meant by specific operation?

We should return to Leibniz for the answer. We have to search for machines which are '... still machines in their smallest details up to infinity...' by limiting 'infinity' to the infinitely small as it was understood at the time of Leibniz. Are there such machines? Are they possible at all?

Undoubtedly, the most spectacular experiments in chemistry are the so-called oscillatory reactions. It is simple to demonstrate them: two solutions of appropriate compositions are mixed in a beaker and stirred. After a few seconds the solution becomes blue, and a few seconds later it loses its colour. This change is periodically repeated until the reactants are consumed. The main point here is not the change of colour, since many different oscillatory systems can be prepared. What is important is that this solution changes its properties periodically, like the pendulum of a clock changing its position. We could even say that it 'ticks' like a clock, and if we wanted, we could even use it for measuring time. Moreover, it is very probable that living systems use this effect to measure their internal time; there are several oscillatory systems among the reaction systems of metabolism.

In order to operate in such a periodically regulated way, the system needs a machine with components ensuring feedback(s), i.e. inhibitions necessary to produce periodic changes (oscillations). If the solution in the beaker oscillates, it must have components which are ordered in a given structural system which makes this operation possible. But where are components in such a solution? And in what order should they be arranged, so that this 'clock' operates even if we shake or stir the solution?

If we halve this 'ticking' mixture and put each half into a separate beaker, both halves operate, i.e. they 'tick' further. Such divisible clocks are unknown among man-made clocks. This halving can be continued to 'infinity', or at least to infinity as defined at the time of Leibniz. Even a microscopic drop of this solution is 'ticking'.

This 'chemical clock', or liquid machine, is a complete machine in each of its details; it is not constructed from parts like a mechanical clock, a radio, or a vacuum cleaner, which cannot be cut into many small clocks, radios, or vacuum cleaners. Thus in the 'chemical clock', i.e. in the oscillatory reaction, we have found a machine which corresponds to Leibniz's criteria for a natural or divine machine, which is also a machine in its smallest details.

Today it is obvious that this division cannot be continued to infinity, but only to the point where the liquid can still be considered continuous, i.e. to a volume in which the number of reacting atoms is large enough to be described by the laws of statistics. At present we know that these natural or divine automata are not machines in their smallest details; the dimensions of a drop of cytoplasm cannot be reduced under the continuum limit without loss of its ability to operate. And we even know now that the lowest limit of 'life' lies at volumes which are orders of magnitude larger, although for other reasons. Is it possible that the 'machines' of life and the oscillatory chemical 'machines' operate on common principles?

Several observations offer prima facie support for this hypothesis. Anyone who has seen an amoeba changing its position under a microscope, or in a film or videorecording, knows that as it moves on the microscopic slide or just flows around its food by extending pseudopodia, the cytoplasm inside it whirls around, swirling at random, and during this whirl everything in the cytoplasm operates regularly in a controlled, moreover program-controlled, fashion. Whirling cytoplasm can also be observed under a microscope in the cells of the water plant *Vallisneria spiralis* during photosynthesis, and in many other cells as well. Cell cytoplasm operates completely normally during this swirling process. Thus the cytoplasm is a machine whose operation is not disturbed by stirring, whose internal organization is not destroyed, and whose regulated character is not ruined by chaotic wandering of its components. Thus cytoplasm[S5] is a fluid automaton, similar to an oscillatory chemical system.

It has already been mentioned that the cytoplasm and the nucleus can only develop together: no multicellular organism can develop either from a nucleus alone or from the cytoplasm of an ovum without a nucleus. This is because the nucleus carries the information concerning development and operation, and the cytoplasm is the machine operating according to this information; thus the two parts together make a complete whole.[G13] We can expand this statement to the cell, since an embryo cannot develop from the ovum if either its cytoplasm or its nucleus is missing, and no cell can operate without these either. More accurately, the simplest so-called prokaryotic cells, which also include bacteria, do not contain a nucleus of a defined shape and structure; instead, they contain a DNA coil carrying hereditary information controlling their cytoplasm.

There are some types of hereditary information without which the given cell is inviable, but there are also some types in whose absence the cell is

[S5]This example is slightly misleading, since cells with a nucleus (eukaryotes) have an internal, albeit dynamic, scaffold—the 'cytoskeleton' which is not present in the simplest cells (bacteria or prokaryotes). So-called mechano-chemical molecules play a crucial role in the activities of the cytoskeleton. The reader should consult Alberts *et al.* (1994) for a modern account of molecular and cell biology.

[G13]The extent to which information is localized in the nucleus is open to debate. Developmental systems theorists argue that information cannot be so localized (Griffiths and Gray 1994; Griffiths and Neumann-Held 1999; Oyama 2000; Oyama *et al.* 2001).

capable of living and operating, albeit not so efficiently as when it is in possession of the complete information set. Genetics and gene technology have methods which can be used to realize this. However, this information set cannot be subdivided. (into halves, quarters, tenths, etc.) i.e. an arbitrary proportion of it cannot be removed without damaging the cell.

In contrast, the cytoplasm can be subdivided, i.e. an arbitrary proportion of it can be removed from the cell without causing cell death, unless this proportion is too large. For example, the German scientist Hartmann cut down about a third of the cytoplasm of an amoeba every second day, and repeated this for 130 days without killing the amoeba. Thus the cytoplasm is really a 'natural automaton' in the sense of Leibniz, which can be divided, and the halves, quarters, tenths, etc. so produced are also automata of full value, similar to oscillatory reactions.

Thus if cytoplasm can be 'halved', i.e. it is a 'natural' automaton, and it is also a fluid automaton, we need only determine whether it is also a chemical automaton. The operation of every machine, device, instrument—every continuously operating system—is based on energy flow. Energy enters the system from somewhere and eventually leaves it. While it is within the system it is manipulated so that part of it is forced to make the system operate, whereas the other part leaves the system, mainly in the form of heat. Man-made operating systems, machines, instruments, and automata manipulate the energy used so that it performs useful work. For example, the potential energy of wind or water in wind- or watermills (or at least part of it) is used as mechanical energy for grinding corn, which otherwise would be very tiring work. Work is also performed on the screen of a television set when providing the image by deflecting the electron beam to and fro, although this work is not directly perceptible to the viewer. Here, electric energy is manipulated in the television set.

Combustion engines are driven by chemical energy. However, this chemical energy is transformed at the moment of explosion into volume energy and, by pushing the piston, it is transformed further to mechanical energy driving the motor or the car.

Living systems are also operated using the chemical energy of nutrients. This statement is also valid for plants, although the nutrients are produced by the plant itself via photosynthesis. In more developed living systems, this chemical energy can also be transformed into mechanical energy (e.g. into movement, effort) or even into electrical energy, not only in electric rays and eels, but also in our nervous system. However, the basic operation, the operation which characterizes life itself, is always of a chemical nature. In the simplest living systems (e.g. mycoplasms or thermoplasms) there is no mechanical operation or electric organs.[S6] Only chemical processes take place in them, and the useful work performed, i.e. the production of their own material for growth and proliferation, is also of a chemical character.

If these living systems operate in a controlled and regulated way such that the entering energy is chemical energy and the useful work is chemical work, and if the chemical energy is not transformed into mechanical or some other type of energy, then the manipulation of

[S6]More accurately, they are directly tied to the action of molecules. For example, one can measure the mechanical force generated by an enzyme copying a piece of DNA. The enzyme has to move along the DNA strands and this movement is propelled by ATP (adenosine triphosphate). One can measure the pulling force of the moving enzyme (Alberts *et al.* 1994).

energy will also occur in a chemical manner. In this case, cytoplasm is not only a fluid automaton and a 'divisible' automaton, but it should also be a chemical automaton, like an oscillatory chemical reaction.

It is easy to see that such a chemical automaton would be a complicated system, even in the case of the smallest, simplest, and most primitive living system. Thus it is not sensible to study the basic operating principles of chemical automata in such complicated systems if there is another simpler possibility. As we have already seen, there is such a system—the oscillatory chemical reaction. Thus before returning to the study of the genesis of life, let us consider the basic operating principles of fluid chemical automata in simpler chemical systems. We shall then apply these principles to the processes of the genesis of life.

2.3 Order in the nothing

The wheels of a clock are arranged in a defined order which can be observed visually. The clock operates only if its wheels fit into this order; it cannot operate properly with any other arrangement. This order allows appropriate manipulation of energy passing through the clock, starting from the spring and going through the forced trajectory created by the wheels to the hands of the clock, thereby performing useful work. The operation of the clock and other mechanical machines is based on directing mechanical energy through forced trajectories. No wheel can move without the movement of another wheel, and the magnitude and direction of movement of one wheel are strictly defined in terms of the movement of the previous one. The translation of mechanical energy via forced trajectories can be seen in these machines, and the operation of a mechanical device is intelligible even to the layman.

Energy also passes through many forced trajectories in electric and electronic instruments, but here the energy is electrical energy, and the forced trajectories are wires, integrated circuits (ICs), silicon chips, and other conducting or semiconducting components. They are also organized in a strictly defined order; the wires providing the forced trajectory cannot be connected arbitrarily, if we want the radio to broadcast or the electromotor to rotate. This order is understood only by experts, but we can all understand that a strictly defined order exists in our television set, radio, or computer.

But what order can exist in a solution which is, in addition, vigorously stirred? What can be ordered in a chemically oscillating liquid, and into what order is it arranged? However, some type of order must exist, as the energy passing through it is manipulated in a regular way. There must be forced trajectories through which the energy passes, and there must be a definite order which creates a forced coupling between the forced trajectories so that the solution is capable of operating as an automaton with a given function.

Chemical reactions also have directions. This is indicated even for a non-chemist by the arrows which are sometimes used instead of the equality sign in chemical equations to indicate that the reaction proceeds from left to right or vice versa. Double arrows pointing in opposite directions are also often used to indicate that the reaction can proceed in both directions depending on the concentration of the reactants. The arrow also indicates the direction of energy flow, since not only the amounts of the substances but also the energy contents on each side of the chemical equations should balance, corresponding to the principle of energy conservation.

But from where to where is energy flowing in these reactions? Whatever method is used to study a solution in which a chemical reaction proceeds, no energy flow can be observed, let alone its direction. Matter and energy are distributed homogeneously and uniformly in solutions, and this homogeneous distribution is preserved during the reaction. What sense can then be attributed to the arrows used by chemists, if they cannot be identified with a geometrical direction?*

This is a difficult question, which we first consider in terms of philosophy. The reductionistic thinking, or attitude, generally adopted in science aims to deduce everything from three basic concepts: matter (including energy), space, and time. We say that matter is something existing objectively, that this existence is realized in space, and that its changes reflect time. We do not allow matter and time to conflict in this respect, but science registers numerous (or perhaps even innumerable) changes to which an unambiguous direction can be ascribed without this direction being identical with any of those of geometrical space. This can be observed in everyday life, (e.g. economic changes) or even the changes in our own bodies. For example, the ageing process is characteristically unidirectional and, unfortunately, cannot be reversed, although its direction cannot be identified in geometrical space.

Mathematics appears to be suitable for solving such scientific problems. In the same way that it calculates the time course of the geometrical position of a body using equations and graphs, it is capable of calculating and plotting changes in a body which do not originate from changes in the geometrical position. Not only quantitative changes, but also the time course can be calculated in this way. Mathematics uses the assumption that there are fields whose size is not measured by length, but by weight, charge, magnetic field strength, chemical potential, or any other calculable and measurable unit, and it examines the course of processes according to their direction and dimension in these imagined abstract fields, the so-called state fields.[G14]

The usefulness of this procedure has mainly been demonstrated in physics. It has been used to calculate changes in the world accurately and quantitatively, and has also been used for prediction with excellent results. These methods have made the development of high technology possible.

[G14]The state-space approach to scientific models has been extensively explored in the philosophical literature on biological science (Beatty 1981; Lloyd 1988) as well as on physical science (van Fraassen 1980; Giere 1988).

*It should be noted that oscillatory reactions in which the 'ticking' is both time and space dependent can also be produced. In these, colour waves move in different directions. However, this is another question which does not contradict our previous discussion of homogeneous oscillatory reactions mixtures.

When a modern engineer uses mathematical equations in his design, he never thinks that he often plans the operation of his instruments in real space by thinking in an imaginary abstract state field.

Chemistry also utilizes abstract state space successfully (mostly in physicochemical calculations), and calculations with many variables applying abstract state space are frequently used in ecology. However, chemistry has developed mainly in terms of individual reaction steps. If one wants to synthesize a complicated compound, it is necessary to perform many transformations on the starting substance to obtain the desired end product, and these steps are performed one by one, by preparing the intermediates and letting them react further. Thus the experience of chemists in the state field mostly involves one-step reactions, and no methods have been developed for investigating many reactions taking place simultaneously in a complex formation–transformation relationship.

Even in the simplest living organism, many thousand chemical reactions take place simultaneously so that each of them is in a direct relationship with at least two (but sometimes three, four, or more) other reactions and, through them, it is indirectly connected to all the reactions, thus forming an extraordinarily complex network. Biochemistry started with the investigation of these complex reaction networks, and it achieved significant qualitative results by discovering what is formed from what and into what the products transform. However, so far it has been unsuccessful in developing quantitative descriptions. The reason for this was that it tried to study these networks quantitatively using chemical kinetics, which is the study of the time course of chemical reactions. This is done using differential equations, but those describing reaction networks are too complicated to handle mathematically, at least in their general form.

This fact poses a severe problem, since the solution depends on mathematical manageability. The end-product of a chemical reaction can be further transformed with new reactants, and the new (second) product can react further to give a third, fourth, fifth, etc. product. Obviously, not only the first reaction, but the second, third, etc., have directions in the chemical state field, and these directions define a forced trajectory through which chemical energy is forced to pass. If we do not consider the time course of the process and restrict ourselves to making a balance, like a chief accountant of an enterprise on the profit at the end of the year, the events can be calculated using a simple arithmetic method. This is called chemical stoichiometry; in popular science its equations are known as chemical equations, and in science they are known as stoichiometric equations.

By means of stoichiometry, the material and energy balance of a chemical forced trajectory can easily be calculated by simple additions, whatever the number of elementary chemical steps in the reaction chain. However, stoichiometry fails if the reaction chain branches, and it will certainly fail if there are stoichiometric feedbacks in the reaction chain.

What is stoichiometric feed-back? It was stated above that chemical reactions indicate a definite direction in the abstract chemical state field. If we allow the end-product of a reaction to react with a new reactant, then this indicates a new different direction. By repeating this process with a succession of new reactants, one can 'wander' in the non-existent abstract field. The direction of the reactions depends on the reactants. There are cases in which the directions of the subsequent reaction pathways change in such a way that eventually we arrive at our starting point, i.e. after several or many chemical transformations we have the same chemical compound with which we started. This can be interpreted as going round a circle in the chemical state field, and this is what is meant by stoichiometric feedback. In chemistry these processes are simply called chemical cycles.

Chemical cycles occur mainly in biochemistry. Moreover, since every enzyme-catalysed reaction is itself a cycle, biochemistry is basically the chemistry of cycles. Such cycles include the Krebs cycle and the Calvin cycle, by which carbon dioxide is transformed into sugars in plants by photosynthesis. Now we can understand why we have said that enzymes 'perform' a chemical transformation, although they are only substances. In other words, we can now understand how it is possible that a substance 'operates'. The enzyme goes round the abstract chemical state field by reacting with other reactants, and while they are transformed, the enzyme transforms back into its original form.

This whole process can be considered from two directions. Let us examine this using the example of a four-stroke petrol engine. This also operates in a cycle, which has four elementary steps or strokes. We have become accustomed to studying this engine from the viewpoint of work, because we use it to do work. Thus we can say that our engine functions with the energy of petrol by emitting mainly carbon dioxide and water vapour (these are the major components of the exhaust gas).

Readers would be very surprised if they were told that the petrol engine is an instrument which accelerates the decomposition of petrol into water and carbon dioxide by a factor of several million. This aspect of the process was revealed by chemistry, when it discovered catalysts (enzymes). According to chemistry, enzymes (catalysts) are substances capable of accelerating the rate of chemical reactions (often by a factor of several million). It cannot state this differently, as it has not recognized the nature of enzymes as chemical engines.

However, this is only a half-truth, since it does not answer the question of how enzymes do this and does not provide a solution to the problem of where the enzyme gets the energy needed for this process. It would be more accurate to say that an enzyme is a chemical engine driven by the substrate (the substance to be transformed) while the substrate itself transforms into another substance (with a lower energy content). The substrate is the petrol of the 'enzyme engine'.

This is a false analogy only in so far as the petrol engine is not a single wheel but a complex machine, whereas the enzyme cycle corresponds to a single 'chemical wheel'. But a complex machine is needed to transform chemical energy into mechanical work. If we keep to the same kind of energy, a simple example can also be obtained from mechanics. A water-wheel does not operate by exchanging the water in the lower position with the water in the higher position. Rather, the water in the higher position drives the water-wheel by transformation to a lower potential energy.

We have said before that biochemistry is basically the chemistry of cyclic processes. This cannot be repeated too many times. It is true in the sense that each individual chemical reaction of metabolism is catalysed by individual specific enzymes, i.e. each chemical transformation takes place by means of a small chemical cycle. However, it is also true in that these enzyme-catalysed reactions are also organized into chemical cycles (e.g. a Krebs cycle) in metabolic processes, and the operation of these cycles is coupled. Cell metabolism is similar to an instrument consisting of many thousands of cog-wheels, only ordered in the chemical rather than the geometrical state field. It could be claimed that although this provides an interesting comparison, the basic principles are totally different. Where are the teeth of the chemical wheels? Chemical cycles are not coupled with teeth, but through the fact that the product of one cycle is the fuel of another cycle.

Whether the basic principles are identical or different could be established if the two types of machines could be described mathematically. However, classical stoichiometry is not suitable for the description of cyclic processes. Fortunately, there is an approach known as cycle stoichiometry which enables cycles and simple networks to be described algebraically, using simple arithmetic operations, and also allows mathematical formulation of the conditions for coupling cycles. Surprisingly, the rules for the coupling of cog-wheels and the coupling of chemical cycles are described by the same mathematical equation.

Of course, this does not mean that chemical cycles are cog-wheels. However, it does mean that, using this approach and thinking in a suitable abstract chemical state space (or, more precisely, in a stoichiometric state space), it is possible to design fluid chemical automata in a similar way to that used by mechanical engineers when planning mechanical automata or by electrical engineers when designing a radio or a computer. The same method is used and the work can be done at a desk.

It is not possible here to go into details concerning this planning; those who are interested can find further details in textbooks or professional journals. The author has designed several 'chemical machines', of varying degrees of complexity, at his desk, and a number of them have been applied in industrial technologies. As an example, only for the purpose of demonstrating the 'flowsheet' of such a fluid machine, the design of such a 'ticking machine' is shown in Fig. 2.1. Figure 2.2 shows a computer simulation of the operation of a chemical machine constructed in this way. Figure 2.3 shows that the elementary steps of the most famous oscillatory chemical system, the Belousov–Zhabotinskii reaction, fit into such a system.

2.4 And who is the constructor?

Wonderful, the reader could say. Science has shown that it can arrange non-existent wheels in an imaginary field into an invisible order, thereby creating an operating machine which really exists. This is indeed a record.

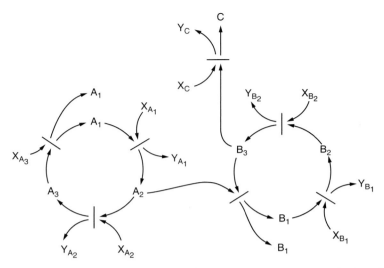

Fig. 2.1 The connection diagram for a fluid machine or, more precisely, a fluid 'ticking construction' designed theoretically by considering the basic properties of the participating chemical reactions.

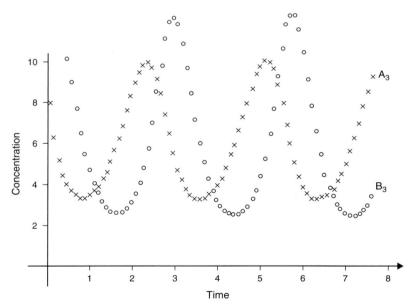

Fig. 2.2 A computer simulation of the operation of the construction shown in Fig. 2.1. The concentrations of the substances in the solution swing to and fro, i.e. oscillate.

It is also a splendid recognition that living systems are complicated fluid machines consisting of invisible wheels in an imaginary field. But if this is true, then those claiming that such a machine could not be developed by itself are probably correct. Following this line of thought, we have

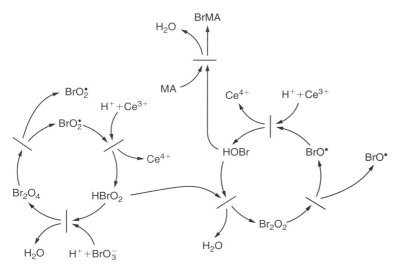

Fig. 2.3 The elementary chemical processes of the most famous oscillatory system (the Belousov–Zhabotinskii reaction) have already been identified. These chemical processes can interact with each other in the manner described by the theoretical scheme shown in Fig. 2.1.

arrived at a point where not only should the genetic program have been born by itself, but the machine controlled by it should also have been constructed by itself. Obviously, a constructor was needed who designed the controlling program first, but also planned the controlled machine. Who was the constructor who designed these congenial machines and who was the chemist who realized these plans?

In order to be able to answer these questions, we should first become acquainted with a particular type of chemical 'wheel'—self-reproducing chemical 'wheel'. Such 'wheels' also exist in chemistry.

Living systems are self-reproducing systems since their basic property is their ability to proliferate. This is trivial. However, the fact that self-reproducing automata can be constructed artificially is not trivial, particularly as such automata have not yet been produced, either in mechanics[S7] or in electronics. However, John von Neumann proved that this is theoretically possible; thus we can hope that sooner or later such automata will be realized. Nevertheless, von Neumann did not propose the existence of automata whose component parts (wheels) could reproduce themselves as this would be impossible with hard automata. Self-reproduction in such machines is prevented by geometrical constraints. Imagine what the consequences would be if a car reproduced itself by the proliferation of its wheels, crankshaft, etc. while it was being driven. However, there are no spatial constraints in the reproduction of chemical 'wheels', and so in chemistry both the self-reproduction of wheels and the reproduction of the machine by the self-reproduction of its components is possible, both theoretically and practically.

[S7]L.S. Penrose and R. Penrose (geneticist father and mathematical physicist son) constructed working self-reproducing mechanical devices (not to be confused with automata) (Penrose 1959). They are carved out of wood, and are able to undergo 'trivial' self-reproduction by the incorporation of neighbouring building blocks. Molecular motion must be implemented by shaking the box containing these machines.

Let us assume that we have a molecule containing two carbon atoms (and several other atoms) which can react with a molecule containing one carbon atom. The product is a molecule with three carbon atoms. In a subsequent reaction this reacts again with a molecule containing one carbon atom, and thus we obtain a molecule with four carbon atoms. If this reacts again with a molecule with one carbon atom, a molecule with five carbon atoms is produced. Let us assume that this molecule is unstable and, as a final reaction step, it can split into a molecule with two carbon atoms and a molecule with three carbon atoms. If the molecule with the two carbon atoms is identical with the original molecule, then we have arrived at our starting point, i.e. we have obtained a chemical cycle since the end-product can pass through the reaction chain described again and again, by producing a molecule with three carbon atoms in each cycle. Thus we can say that the cycle is a chemical machine capable of the continuous production of molecules with three carbon atoms from molecules with one carbon atom.

But what happens if the molecule with four carbon atoms in this reaction series is unstable and splits into two molecules with two carbon atoms (identical with the starting molecule) before it can react with a new molecule with one carbon atom? In this case we have a machine which, on the one hand, produces molecules with two carbon atoms, and, on the other hand, transforms itself back into a molecule with two carbon atoms. This means that at the end of the cycle we have produced two of our starting molecules. Now, both can react further; after another cycle there will be four molecules with two carbon atoms, after three cycles there will be eight, then 16, 32, 64, 128, etc., i.e. this molecule, which can go through a cyclic reaction, proliferates.

Adenosine triphosphate (ATP), the general energy carrier of organisms, is produced in such a self-reproducing process, which is known as an autocatalytic process. Moreover, the cytoplasm itself is also an autocatalytic system. No living being can operate without autocatalytic cycles; this is the capital which can be made to earn interest.

Thus self-reproducing chemical wheels do exist. Not only is the synthesis of ATP such a system, but so is the Calvin cycle. Moreover, the Krebs cycle has a self-reproducing variant operating in the opposite direction, which is called reductive citric acid cycle. These are obviously not as straight forward as the simple illustrative example discussed above; the molecules have many carbon atoms and are quite complicated, and the cycles have more than three or four steps. In addition, every step of the ATP, Calvin, and reductive citric acid cycles is an enzyme-catalysed reaction. However, in the genesis of life, appropriate enzymes were unlikely to be available for the formation of self-reproducing 'wheels'. Do we know of any autocatalytic cycles with steps which are not enzyme-catalysed reactions?

Let us now examine the two wheels of the 'ticking chemical machine' described at the end of the previous section in more detail. In Fig. 2.1, which shows the processes in an abstract manner with the compounds identified by letters, component A_1 in the first cycle and component B_1

in the second cycle are duplicated during one turn of the cycle, i.e. both cycles are autocatalytic. This fluid 'ticking' system is a machine with self-reproducing wheels. In the language of chemistry, this oscillatory chemical system is obtained by an ingenious coupling of two autocatalytic cycles.

The same result is obtained in Fig. 2.3. Here, BrO_2 is duplicated in the first cycle and $BrO\cdot$ is duplicated in the second cycle. These molecules can then react further, successively duplicating the amount of material.

However, if this is the case, the amounts of material should increase exponentially and not oscillate, but Fig. 2.2 shows oscillation. Well, this construction is ingenious because it converts the exponential increase into an oscillation, just as a pendulum clock does, only by a different mechanism. The pendulum transforms the exponentially increasing rate of free fall into an oscillatory movement, whereas the chemical machine achieves this transformation by 'feeding' the second cycle with material from the first cycle while, at the same time, material is consumed by chemical tapping leading to BrMA (bromine malonate). This construction is similar to the predator–prey relationship between foxes and rabbits. The exponential proliferation of rabbits is constrained by the proliferation of foxes, which is itself constrained by the fact that if the fox population becomes too large they starve to death. At the beginning of the twentieth century, Lotka and Volterra showed that such a two-component predator–prey relationship leads to an oscillatory change in population size.

Several examples of autocatalytic cycles have been encountered in which organic compounds are formed, but they are extremely complicated. We have shown that the elementary steps of the Belousov–Zhabotinskii reaction can be organized into simple autocatalytic cycles, but the components of this system are inorganic compounds. Are there any organic compounds capable of producing simple autocatalytic cycles with only a few steps?

In fact such a reaction does exist, which is similar to the simple illustration discussed above, producing molecules of four carbon atoms from molecules of two carbon atoms which are recovered by the splitting of the product. This reaction is known as the formose reaction. The starting molecule is glycol aldehyde which contains two carbon atoms. This reacts with formaldehyde, which contains one carbon atom, to form glycerol aldehyde which contains three carbon atoms. This in turn reacts with formaldehyde, producing a sugar molecule containing four carbon atoms, which splits into two glycol aldehyde molecules, each containing two carbon atoms, and the process starts again (Fig. 2.4.).

The formose reaction has been known for over a century. However, in practice it is not so simple as numerous side-reactions are possible. For example, the sugar containing four carbon atoms can also transform into a sugar containing five carbon atoms and so on, and these can also be fed back into the basic reaction. Thus a very complicated reaction network can be developed which is also autocatalytic overall, which is the most important point of interest for our purposes. A series of patents exist for

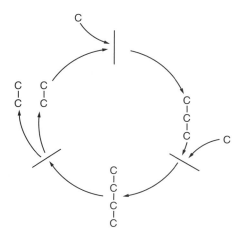

Fig. 2.4 The most promising potential of fluid machines is that self-reproducing parts can also be constructed from their components, and that self-reproducing, and even proliferating, automata can be built from these parts. Living systems grow and proliferate in this way. The figure shows the operating principle of the simplest 'self-reproducing chemical wheel' involving an organic compound containing two carbon atoms (glycolaldehyde). This reacts with a molecule containing one carbon atom (formaldehyde) to form a molecule containing three carbon atoms (glycerolaldehyde) from which, in a further reaction, a compound with four carbon atoms (tetrose) is obtained. This molecule splits into two molecules each containing two carbon atoms (glycolaldehyde), thus reproducing the starting molecule. The process can continue with both molecules resulting, in the production of 4, 8, 16, 32, 64, ... molecules of glycolaldehyde in subsequent cycles. For simplicity only the carbon skeletons of the molecules are shown.

the industrial production of sugars from formaldehyde using the formose reaction.

Concerning the role of the formose reaction in living systems, the fact that it builds a complex rather than a simple network is an advantage rather than a disadvantage. In the formose reaction, all kinds of sugars may be formed or even transformed into each other. The basis of the metabolic network of living systems is just such a reaction network formed by the transformations of different sugars into each other. However, these transformations do not occur directly, but via phosphate derivatives mediated with ATP as mentioned earlier.

Research into chemical evolution has been unable to prove that ATP was present on the primordial Earth before the appearance of life. Thus the transformation of sugars in the metabolism of the first living organisms could not take place via phosphate derivatives. Is it possible that the chemical steps of the formose reaction served as the basis for the metabolic network of primordial living organisms? Is it possible that the materials of the formose reaction network constituted the capital whose interest later led to life?

Formaldehyde is one of the most common organic compounds in the Universe. It was also present in the atmosphere of primordial Earth.

Moreover, it was not only present, but was continuously produced from atmospheric methane by ultraviolet solar radiation, lightning, radioactive and cosmic irradiation, etc. One molecule of glycol aldehyde molecule is formed from two molecules of formaldehyde, and this reaction, although very slow, does takes place.[S8] The further reactions of glycol aldehyde with formaldehyde are sufficiently fast that it is likely that the formose reaction must have been one of the most frequent chemical processes on primordial Earth. And it is probably one of the most frequent systems or processes everywhere in the Universe where liquid water is present. Thus the 'self-reproducing wheel' did not have to be designed; neither a constructor applying mathematics nor a chemist working with laboratory flasks was needed to produce them. Nature was the constructor, and the planets served as flasks during their formation. Thus, under suitable conditions, self-reproducing wheels can be developed without any help from humans or other intelligent beings. This theory is supported by a variety of experiments concerning the spontaneous formation of organic compounds under primoridal conditions, i.e. chemical evolution, carried out in different laboratories all over the world. In these chemical evolution experiments the question of which chemical reaction systems and which fluid machines operate was not of interest; rather, the products formed were studied. Nevertheless, the fact that many different and very complex compounds are formed from a few simple basic compounds in these reaction mixtures proves that many different chemical processes have to take place simultaneously and in interaction with each other. The only question is whether these spontaneously formed chemical machines have been capable of functioning as the machines of living systems.

There is a sugar containing five carbon atoms which is called ribose. This is also produced in the formose reaction, and is the compound to which the nucleotide base, serving as a letter mark, is coupled in ribonucleic acids. Thus the formose reaction is a suitable machine for the production of the raw materials for the self-reproduction of the controlling unit. Another intermediate of the formose reaction is glycerol aldehyde, from which glycerole, which is one of the raw materials of membranes, is produced. Thus the formose reaction also provides the raw materials for biological membranes. Thus this reaction system appears to be a suitable source for the spontaneous development of living systems under the conditions existing on primordial Earth. However, in order to be sure of this, we need to know the basic construction of living systems. We need the design—the flowchart—of the simplest fluid machine about which it can correctly be stated that it lives. Only then can we investigate the suitability of the formose reaction for functioning as part of such a machine.

[S8]Albert Eschenmoser and his colleagues (Müller *et al.* 1990) have shown that if glycolaldehyde is replaced by glycolaldehyde phospate, the diversity of products is reduced and ribose phosphate is generated in a considerably higher relative amount than ribose in the original system. This is important because ribose is a major building block of RNA (ribonucleic acid), which is believed to have played a very important role in early evolution (see later).

2.5 Chemotons

The reader could claim that it is reasonable to accept that such fluid machines could develop in the primordial soup, even if they were self-reproducing. After all, they are only chemical processes and they still operate in the natural world, although they are no longer self-reproducing. However, we have stated that living systems are program-controlled chemical machines. So where is the program? It can be conceived that in self-regulation takes place chemical processes, and that in a complex chemical reaction network these processes are regulated by feedbacks. However, for program control, external intervention is necessary which regulates, via external information, the operation of the machine Crick (1981) has not claimed that a machine cannot develop, only that the probability of the spontaneous formation of a program is low.

This is true if, by the term program, we mean a set of laws, a coded text on a tape or a disc, or a complicated system of holes on a punched card. However, early technological development of controlled machines did not involve complicated coded texts, but very simple primitive constructions. Why do we think that life has developed in the opposite way, by starting with extremely complex constructions?

For a deeper understanding of this question, let us consider first a simple program-controlled mechanical construction: a vessel with a nozzle at the bottom into which water drops slowly. This system itself is not program controlled, since as water drops into it at the top it simultaneously flows out at the bottom. It would be program-controlled if these was a stopcock at the bottom which was programmed to open only when the prescribed amount of water had been collected and then to close again. Of course, this can easily be realized by connecting thin rubber tubing to the outlet, leading it up to the desired height, and then turning it down below the vessel. Thus, initially the water cannot flow out; its level rises in the vessel and also in the tube, up to the turn in the tube. Then the water starts to flow down the tube and sucks the water from the vessel itself, in a similar manner to drawing off wine from a barrel using a rubber tube (Fig. 2.5). When the vessel is empty, the process starts again from the beginning. The amount of water flowing out of the vessel can be 'programmed' by adjusting the height of the turn in the rubber tube.

If something similar could be realized in chemistry, we could obtain a primitive program-controlled system which did not require coded information. In this system, rather than water, a solution of some compound would be collected in the vessel and the outflow would not take place through a rubber tube, but in a chemical manner, via a reaction of the compound. The reaction should be such that it could not 'flow back', i.e. the product could not transform back to the original compound. For example, the molecules of the original compound should combine strongly to form an unbreakable molecule chain, i.e. polymerize. Thus our program-controlled system contains a chemical 'machine' producing a compound capable of polymerization (a monomer) and a mechanism ensuring that this monomer would not polymerize until it has reached a desired concentration. Once this concentration has been reached, polymerization should start by consuming, or 'sucking away',

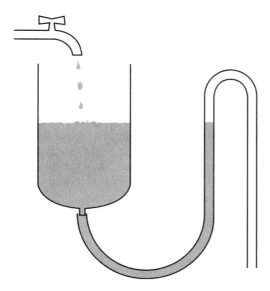

Fig. 2.5 Program control is always assumed to be electronic, although mechanical program control solutions were widely realized before the development of electronics. Program control can also be carried out in a very primitive way. In the system shown in the figure, the amount of water sucked down can be 'programmed' by the height of the turn in the siphon. A similar system can be realized in fluid machines by an intermittent 'sucking down' of the increase in the concentration of the components by a chemical method.

the monomers down to the original low concentration, after which the machine begins to 'fill up' again.

However, the monomer starts to polymerize when it wants and not according to some program. This is correct, except for template polymerization in which the monomers do not bind to each other at all, or bind only at a very low rate, unless a 'pattern' molecule is present on which the monomers can be arranged in a line and combined at a high rate. Thus template polymerization cannot start until the template molecule is added to the monomer solution. We have encountered a similar process earlier in this chapter—the synthesis of nucleic acids occurs in the same way. Nucleic acids are polymers consisting of two helical strands. Each strand, when separated, is capable of acting as a template on which a new strand can be formed by polymerization. The double strand cannot serve as a template; only the individual strands have this ability. At high concentrations of the monomer the double-strand structure tends to break up, promoting the separation of the two strands. Thus it is a candidate for use in a simple program-controlled fluid machine.

We shall consider a chemical cycle (network) producing monomers capable of template polymerization, and a template constructed from polymers and having a double-strand structure. The cycle produces

[S9]Self-replication of *long* templates is the Holy Grail of the field of chemical evolution. *Short* templates can self-replicate easily because the two strands dissociate from each other with a certain probability and thus can serve as templates again (von Kiedrowski 1986). Longer strands also can be efficient templates, but there is no replication because the two strands do not separate spontaneously: some intervention is required. Three solutions have been proposed: (i) a surface helps to promote strand separation (Luther *et al.* 1998); (ii) the template could act as a replicase enzyme (not yet achieved); (iii) the special periodicity in monomer concentration outlined by Gánti in the text (not yet tested).

[S10]There are some simplifications in the theoretical treatment which make the conclusions less firm than stated in the text. However, it is noteworthy that in the first edition of *The Principle of Life* (in Hungarian, 1971) this system was called the chemoton, and was modified later to include the membrane subsystem. Two chemists attacked the problem of the origin of life in 1971: Manfred Eigen and Tibor Gánti. The primary approach in chemistry is based on well-stirred spatially homogeneous systems. It is remarkable that both of them missed the crucial point on spatial separation then.

the monomers, but they cannot polymerize as the template is in its double-stranded state, i.e. it is present in a covered form. Therefore the number of monomers (and thus their concentration) increases. This phase corresponds to the rise of the water level in our mechanical example. However, once the monomer concentration reaches the level at which the strands are separated, polymerization starts and new strands are developed on the individual separated strands. During this process, the concentration of monomers decreases, since some of them are consumed for the synthesis of the new strands. This process corresponds to the sucking down of water from the vessel. When the new strand is complete, the old and new strands again form a double strand, which can no longer act as a template. Thus monomer concentration increases again, i.e. vessel fills up, until it reaches the level when the double strands can separate, and the process starts again.

Thus we have succeeded in outlining a fluid chemical machine, which, analogously to the siphon, operates with a primitive program control.[S9] The difference in the principle of the control is that in the siphon system the filling up is controlled, whereas in the chemical system described above the sucking down is regulated. The concentration needed for the separation of the template is constant for every individual template, but the consumption of the monomer in individual polymerization periods (i.e. the 'sucking down' of the monomer concentration) depends on the number of template molecules present and on the length of strands, i.e. on the number of monomers needed to synthesize a new strand.

Our construction is not yet perfect. In the polymerization period a new strand forms on each template strand, i.e. the number of templates is doubled. Thus the system uses up twice the amount of monomer for polymerization in the second cycle, four times as much in the third etc. In order to ensure an unperturbed operation of the system, the amount of monomer formed should also be doubled in each cycle. Based on our present knowledge, we can suggest a solution for this problem: an autocatalytic cycle rather than a simple cycle is needed for the production of monomers. In this case we obtain a program-controlled fluid machine, with self-reproducing components, which also reproduces itself, cycle by cycle.

Like the oscillatory fluid machine shown in Fig. 2.1, a theoretical 'wiring diagram' of this program-controlled self-reproducing chemical machine can be designed by showing the elementary reaction steps. Cycle stoichiometry (mentioned above) can be used to calculate the exact accurate stoichiometry and construction of the system. Moreover, the complete system can be described, both qualitatively and quantitatively, by a single overall equation for the cycle. Once the elementary reaction steps are known, the operation of the system can be studied using the methods of chemical kinetics and an exact computer simulation of the behaviour of this theoretical construction can be performed. These calculations have proved that such systems do indeed operate in a program-controlled manner.[S10]

However, there are still some problems with this system. We state that it is self-reproducing, but this self-reproduction cannot be observed directly, since a chemist performing the appropriate experiments sees the concentration of the components of the system increasing periodically at an ever-increasing rate. Living systems can also be regarded as self-reproducing program-controlled systems, but their self-reproduction is also reflected in spatial separation, i.e. living beings also proliferate. If we want to proceed further towards living systems, a membrane pouch should also be present in the program-controlled chemical machine. When the material and the program are doubled inside the membrane pouch, the pouch should also be increased and split into two identical enclosed program-controlled machines.

The reader may well think that this is too difficult a problem to solve. So far we have theorized in the abstract chemical state field, without any geometrical limitation. Now we are about to place the operation of this self-reproducing machine, constructed in this abstract chemical state field and operating in a regular controlled way, into geometrical space in order to provide visible evidence of its regulated operation so that our solution, so far studied indirectly by chemical methods, will become an increasing, dividing, and proliferating fluid automaton in the strictest sense of the word.

The geometrical solution for this is suggested by nature, and it is not very complicated. The secret is hidden in the properties of two-dimensional liquids. Many may not have heard of these liquids, although almost everybody has experienced them. Most people have played with soap bubbles. The skin of a soap bubble can be considered as a two-dimensional liquid. Under an external mechanical force it behaves like an elastic solid body, but at the same time, if we observe its coloration we can see the swirling turbulence in the plane of the skin, i.e. it behaves as a liquid. A similar skin, or membrane, covers the surface of all living cells, only this membrane is much thinner than the skin of the soap bubble. It consists of just two molecular layers opposite to each other. Therefore the cell membrane is not expandable like the much thicker soap bubble membrane.

Double layers with the properties of two-dimensional liquids, known as molecular membranes, form spontaneously in aqueous solutions of elongated molecules with one end hydrophylic and the other end hydrophobic. These molecular membranes usually form microscopic spheres, called vesicles, during their development. In chemical evolution experiments they are often formed spontaneously under the conditions modelling those of primordial Earth, but they are known by different names depending on the conditions of their formation: microspheres, marigranules, spherules, jeewanus, etc. We shall call them microspheres.

These microspheres are capable of increasing. If membrane-forming molecules are present, or are developing, in a solution, they are quickly incorporated into the membrane. Each incorporated molecule increases the surface area of the membrane. Small molecules can easily diffuse through these two-dimensional liquid membranes, whereas larger ones diffuse only with difficulty or not at all. If the larger molecule is charged, it cannot penetrate the membrane at all.

Some of the raw materials of the autocatalytic networks mentioned earlier, are simple small molecules, but the intermediates of the network are quite large and may also be charged. For example, the formose cycle is fed by tiny formaldehyde molecules (HCHO) and it produces much larger sugars, the Calvin cycle consumes carbon dioxide (CO_2) and produces sugar phosphates with a strong electric charge, and the reductive citric acid cycle (the reverse Krebs cycle) also consumes CO_2 and forms larger organic acid molecules which have a strong electric charge.

If such an autocatalytic cycle producing compounds with a high molecular weight operates in the liquid enclosed in a microsphere, then these materials will accumulate in it, i.e. their concentration increases since they can permeate the membrane only with difficulty or not at all whereas the small molecules of their raw materials can diffuse easily and quickly into the microsphere from the external space. As the concentration of these large molecules increases, the osmotic pressure relative to their environment increases proportionally. Since, unlike the skin of a soap bubble, the molecular double layer is inexpandable, the microsphere bursts when the osmotic pressure reaches a critical value.

The situation is different if the autocatalytic network in the microsphere, in addition to its own material, also produces membrane-forming molecules. In this case, although the amount of material enclosed in the microsphere increases, the membrane also grows due to the incorporation of membrane-forming molecules produced by the network into its surface. Thus the volume of the microsphere increases. Subsequently events depend on whether the production of the internal materials or the increase in the volume of the microsphere is faster.

Using relatively simple stoichiometric constraints, it can be shown that the numbers of the molecules of the internal materials and those of the membrane-forming material are identical. This means, for example, that when the internal contents of a microsphere of appropriate structure in a starting state of osmotic equilibrium with its environment are doubled, its surface area (*not* its volume) is also doubled. The volume increases much more as, from elementary geometry, the surface of a sphere is proportional to the square of its radius whereas the volume is proportional to the radius cubed.[S11]

Consequently, the unexpected result is obtained that, although the amount of internal material increases continuously, even exponentially, the liquid in the microsphere is diluted (concentrations decrease) because the volume increases more rapidly than the content. If the concentration in the sphere is lower than that in the environment, a relative osmotic vacuum develops in the sphere, and it loses water and becomes dented. However, two-dimensional liquid membranes are only mechanically stable when they are spherical and hence they strive to regain sphericity. This can only be done when both the contents and the surface area are doubled (this can be proved mathematically). Then they can separate into

[S11]This explanation of vesicle division was first proposed by Rashevsky (1938). More detailed calculations (partially based on continuum mechanics) have now been performed showing that the process is feasible (Tarumi and Schwegler 1987).

two spheres, identical with each other and with the original sphere. Obviously, if the concentration remains the same, separation into two spheres means twice the amount of material, twice the surface area, and twice the volume.

Clearly, the processes start again from the beginning in each of the new spheres, the amount of internal material increases, the surface grows, an osmotic vacuum develops, and the two spheres divide again, producing four identical microspheres which are also identical with the original sphere. In subsequent divisions 8, 16, 32, 64, 128, 256, etc. new spheres are formed, are identical with each other and with the original sphere. Hence, the microspheres proliferate in a spatial sense. Thus we have succeeded in transforming the operation of a machine constructed in an abstract chemical field into an event taking place in real geometrical space. We have found a method enabling not only a self-reproducing but also a proliferating fluid automaton to be designed.

If we gave such a proliferating microsphere to a microbiologist and asked him to determine to which species of bacteria it belongs (providing information about the medium as well), after several weeks he would say that it is probably an unknown micro-organism of a similar size to the cocci but, since it has no cell wall, is more likely to be related to mycoplasms or thermoplasms. However, he would be unlikely to guess that what he has 'cultured' on a special medium is not even a living being. He would not think that because the microsphere metabolizes, grows, and proliferates.

Nevertheless, these proliferating microspheres are not yet living beings. Our microbiologist would discover this if he were to use them in genetic experiments. It would be found that no hereditary changes can be caused in them, since they do not have any genetic material—no coded program.[G15] Nevertheless, they have hereditary properties, since we have said that the descendants are identical with the original microsphere.[S12] In this case, there are hereditary properties but there are no hereditary changes. These are hereditary properties which are not mutable properties. These are properties excluded from the range of genetics. They are not carried by some genetic apparatus, but by the construction itself. We have said that a microbiologist would recognize them as living beings in a superficial study, although they are not living beings since they do not have program control. They have no program which could be changed, thus also changing the properties of the system. If they had such a program controlling their operation, then they would also have changeable hereditary properties.

However, we have already designed a program-controlled self-reproducing chemical system. All we have to do is exchange the autocatalytic network of the microsphere for a program-controlled autocatalytic network which is also capable of synthesizing membrane-forming molecules. By ingenious stoichiometric interlocking it can even be arranged that the membrane will not grow when the template molecules carrying the program are not reproduced, but subsequently membrane growth and the division of the microsphere should happen

[G15]This is only true in the narrowest sense of heredity as correlation between parents and offspring due to genes. In the broadest sense of the term, any correlation between parents and offspring is heredity and any change in the correlation is a hereditary change. So-called epigenetic changes in metabolic stable states including metabolite concentrations, structural changes (e.g. in cortical structures), and epigenetic modifications to genetic material all count as hereditary changes in the broadest sense (Jablonka and Lamb 1995; Jablonka and Szathmáry 1995). On the distinction between broad and narrow senses of the heredity relation, see Griesemer (2000a).

[S12]Evolution by natural selection requires a population of evolutionary units which multiply, exhibit heredity, and produce variants. Heredity means that there are units of different types (A, B, C, etc.) and each produces its own type during multiplication. Variability means that heredity is not perfect. It is interesting to consider whether the metabolic and membrane subsystems of the chemoton could produce hereditary variants. In principle they could. Wächtershäuser (1988, 1992) has suggested that certain archaic variants of the reductive citric acid cycle could qualify as units of evolution. This suggestion should be tested experimentally. Cavalier-Smith (1995) introduced the concept of genetic membranes, i.e. those of bacteria, plastids, and mitochondria, the endoplasmic reticulum in eukaryotic cells, etc. Each of them is autocatalytic, by virtue of the fact that they contain special receptor proteins which attract cognate proteins into the

(contd.)

pre-existing membrane. But receptor proteins must also recognize their own type, hence the autocatalytic element. Segré *et al.* (2001) have recently suggested that genetic membranes could have played a role in early evolution ('the lipid world') without the aid of protein enzymes. This suggestion is also hypothetical but testable.

rapidly. The systems thus obtained metabolize, grow, proliferate, and operate in a program-controlled way and are capable of hereditary changes (despite the fact that they do not yet contain any sequence-coded hereditary information). The systems organized in this way are called chemotons.

The connection diagram of a chemoton can be designed in the same way as for the oscillatory reaction system shown in Figs. 2.1 and 2.3. By using this diagram (Fig. 1.1), stoichiometric material balance equations of the processes can be written using cycle stoichiometry, and the exact overall equation of the chemoton, including all its individual components, can be obtained by summing these balance equations, i.e. the organization of the system can be calculated quantitatively. If the rate constants of the individual elementary reactions are known, a differential equation system giving a detailed description of the processes in the system can be derived which, after numerical integration using a computer, will provide the time course of the processes in the system under given chemical and physical conditions.

It is not possible to give a detailed description of this procedure here. Those interested in the mathematical and chemical details are referred to the monograph by Gánti entitled *Chemoton Theory. I: Theoretical Foundation of Fluid Automata.* Further details accessible to non-experts are given in Chapter 3 of this book.

2.6   Chemotons of primordial Earth

Although it is not possible to give a detailed description of the technical construction and operation of chemotons in this book, we can survey their operation schematically and we can also consider whether such proliferating program-controlled systems could have formed spontaneously on primordial Earth or on other planets in which surface water was present. This is important, since if living systems appeared via such chemoton-like systems, then the controlling and decisive role was not chance but chemical inevitability and in this case an exact calculation of the process of development of life becomes possible. It would also allow this process to be reproduced by predictable exact experiments, rather than by modelling based mainly on chance.

First, let us recall the content of a chemoton. It includes a self-reproducing chemical machine producing larger complicated molecules from small molecules with a high energy content. The molecules produced must include molecules identical with the material of the machine itself (this is why they are autocatalytic and self-reproducing). Molecules capable of polymerization on a template, and molecules which (after some transformation) allow membrane growth by incorporation into it. This fluid machine producing chemical substances is the first subunit of the chemoton.

The second subunit is the controlling subunit. This contains a large number of polymer molecules with a double-stranded structure which, at a given monomer concentration, are capable of separating into individual strands, each of which serves as a template for polymerization. (In fact, polycondensation takes place on the template and the byproducts react further with a product of the autocatalytic cycle to give the membrane-forming molecule.) The consumption of monomers in a polymerization cycle (and thus the amount of condensation product formed) is unambiguously determined by the number and length of the template molecules.

The third subunit is a two-dimensional liquid membrane, which forms a spherical enclosure round the system. The small molecules of the nutrients and waste materials can pass through this membrane, but the internal materials of the system are retained.[S13] The membrane-forming molecules produced by the system are spontaneously incorporated into the membrane, thus increasing its surface area.

At this point we should digress slightly and consider the design of the simplest living beings, mycoplasms and thermoplasms. These organisms contain cytoplasm in which, by means of autocatalytic networks, new materials are synthesized from nutrients, from the materials of the autocatalytic network itself, from the materials of the program-controlling genetic subsystem (the raw materials of nucleic acids and proteins), and from membrane-forming materials. They also contain a control subsystem, consisting of double-stranded nucleic acids containing the program and proteins ensuring its implementation, and a two-dimensional liquid membrane which allows nutrients and waste materials to pass through it but retains the internal materials.

Thus chemotons and myco- and thermoplasms (which are unequivocally accepted as living organisms) have identical designs; the only difference is in the technical realization of control. In the simplest chemotons, hereditary information is carried by the number of 'signs' (monomers) from which the template molecules are composed and control occurs stoichiometrically through the regulation of the amount of material produced, whereas in myco- and thermoplasms the sequence of the signs composing the template carries the information and control is achieved by means of protein enzymes via the regulation of the rate of chemical processes, i.e. in a catalytic way. Hence the functional organization of chemotons and myco- and thermoplasms is identical, the only difference being in the technical realization of their process control and regulation.

We can now examine how such a chemoton operates. Let us start from the moment of division, when a new chemoton has already assumed its proper spherical shape. It is then in an osmotic equilibrium with its environment, and nutrients are continuously diffusing into it through the membrane. This is quite a fast process, since nutrients have to penetrate only a distance of few ten-thousandths of a millimetre.

As soon as the nutrient molecules have penetrated the membrane, they react with the materials of the autocatalytic reaction network. Thus internal materials are continuosly produced from nutrients, i.e. the material of the network itself, the precursors of the membrane-forming molecules and the raw materials of the controlling templates (the

[S13]This condition is stringent. A small number of raw materials must be able to pass through the membrane. However, the transport of various ions, necessary for the functioning of any realistic system, presents special difficulties. One can reduce the internal complexity of metabolism at the expense of an increased number of molecules which must somehow be taken up through the membrane from the milieu (Chapter 4).

monomers). However, these monomers cannot polymerize, since the templates are still in the double-stranded form and cover each other's surfaces. Thus the monomers accumulate and their concentration rises. Similarly, the concentration of the precursors of the membrane-forming material also increases, because they are unable to build membranes and hence the membrane surface does not grow.

When the monomer concentration reaches the value at which the double-strand structure becomes unstable, the strands of the template are separated and polymerization (or, more precisely, polycondensation) of the monomers starts on each strand. New polymeric strands are built on the template strands and most of the monomer is used up, so that the monomer concentration decreases. The byproduct molecules formed in the polycondensation process react with the precursors of the membrane-forming molecules, transforming them into true membrane-forming molecules. These are then incorporated into the membrane, which starts to grow and the volume enclosed by it also becomes larger. Hence an osmotic vacuum develops and the membrane sphere is elongated. A neck forms in the middle, and the sphere divides into two identical spheres each containing half the molecules of the catalytic cycle and half the template molecules. Then the process starts again from the beginning.

Once an accurate connection diagram for the chemotons, i.e. their metabolic map, has been designed, chemical kinetic equations describing the time course of every individual reaction can be derived. Thus we obtain an equation system consisting a number of differential equations which can be integrated numerically for individual cases using a computer and this provides a description of the behaviour of chemotons under a given condition. Békés and coworkers at the Technical University in Budapest and Csendes at the A. József University in Szeged have performed such simulation calculations. These studies have proved unequivocally that chemotons work in the way we have shown in our logical deduction, and that this behaviour corresponds to that of living systems.

Obviously, these chemotons are 'only' abstract constructions whose connection diagrams contain letters rather than real compounds, as in the oscillatory system in Fig. 2.1. However, these letters can be replaced by compounds, at least theoretically, as shown in the oscillatory system in Fig. 2.3. The word 'only' is given in quotation marks in order to prevent its interpretation as a depreciation: setting up the abstract constructions is the decisive design step. Similarly, every man-made machine, instrument, or product first takes shape as an abstract construction, although it may not appear on paper but remain in the brain of the inventor as a mere idea. The birth of an abstract construction is invention. When patented, the patent also covers this abstract construction, which is then illustrated by realization examples.

Hence, an abstract construction is essential to man-made inventions, but it must also be realized as a practical instrument if it is to be of any use. The situation is the same here. The problem is whether real compounds can be substituted for the letters in the abstract metabolic design satisfying the conditions of the construction, i.e. whether a plan containing specific compounds can be produced similarly to the one in Fig. 2.3 corresponding to the abstract construction in Fig. 2.1.

In order to investigate the spontaneous pathways leading to the genesis of life, we have to pose this question even more rigorously. Is it possible to prepare a plan—a metabolic map—of a chemoton in which, compounds present in the primordial atmosphere can be substituted for the letters in the abstract reactions? If so, then this will represent a credible pathway for the genesis of life which is not based on chance. If such a design can be realized, this will show that the spontaneous genesis of life cannot be regarded as an accidental improbable miracle, but as a process directed by the laws of nature and taking place whenever and wherever the compounds participating in the design are present and the reactions can proceed.

A solution to this problem is provided by applying the arguments used in previous section. For the following reasons, it is expedient to choose the formose reaction as the first subsystem.

1. Its feedstock, formaldehyde, which is one of the most abundant compounds in the Universe, is believed not only to have been present in the primordial atmosphere, but also to have been continuously reproduced from methane and water as a result of solar radiation.
2. The formose reaction is an autocatalytic network. Thus it is stoichiometrically and kinetically suitable to be a subsystem of a chemoton.
3. The formose reaction produces a large number of sugars, the intermediates in its reactions are mainly sugars, and the basis of the metabolic network of living organisms on Earth is, without exception, the transformation of sugars into each other (today, in the form of sugar phosphates).
4. One of the intermediates of the formose reaction is ribose, a sugar consisting of five carbon atoms, which is one of the components of nucleic acid synthesis and which makes stoichiometric coupling between the operation of the formose network and nucleic acid synthesis possible.
5. Another intermediate is glycerol aldehyde, which can readily transform into glycerol, the basic building block of living membranes. This makes stoichiometric coupling of the formose reaction and membrane formation possible.

Thus the formose reaction seems to be a promising pathway for the process of chemical evolution towards the genesis of life, and a possible basis for the formation of the metabolic network of living systems.

On the basis of the arguments presented earlier in this chapter, it is clear that the information subsystem should consist of double-stranded macromolecules with template properties, together with their replication processes. Experts dealing with the genesis of life and chemical evolution agree that the primordial genetic material was RNA, (ribonucleic acid), and not DNA (deoxyribonucleic acid).

[G16]Recently, Bartel and colleagues have demonstrated the artificial selection of an RNA-based RNA-polymerase from an RNA-ligase, bringing science a step closer to understanding the evolution of the RNA world in which both the catalytic functions of proteins and the information storage functions of nucleic acids are performed by RNA (Johnstone *et al.* 2001; Strobel 2001).

[S14]There is a major difficulty associated with this picture, namely the assumption that a complete network like the one outlined could be realized by a non-enzymatic system. Many would argue that this is impossible. One problem is *chemical incompatibility,* which also arises with chemical evolution experiments of the Miller type. It is true that the synthesis of many interesting compounds proved to be feasible under presumed and simulated prebiotic conditions, but in many cases these conditions are not really compatible with each other. To make matters worse, a number of interesting compounds, synthesized under various conditions, disappear in unwanted side-reactions (e.g. by forming tars) rather than forming more complex compounds of biological significance (see Shapiro (1986) for a harsh criticism of Miller-type experiments). (In fact, one could claim that the most important role of enzymes is to speed up organized reactions *relative to* unwanted side-reactions.) Wächtershäuser (1988, 1992) argued that the surface of pyrite could solve the problem. In his theory, all intermediate metabolism is derived from an expanding network, the constituents of which are bound to pyrite
(contd.)

The construction of RNA is almost identical with that of DNA; the main difference is that the sugar component is ribose rather than deoxyribose. Another difference is that the four bases forming the backbone of RNA are adenine (A), uracil (U), guanine (G), and cytosine (C), whereas in DNA uracil is replaced by thymine (T). Many experiments have shown that the purine bases (A, G) can form spontaneously under primordial conditions. However, there are reports suggesting that the formation of pyrimidine bases (U, C) is also possible; they are probably synthesized on ribose.

Thus the raw materials are available. The question now is whether template polymerization, i.e. the formation of RNA on the template surface, could have taken place in the conditions of primordial Earth. The process is thermodynamically possible, since it proceeds continuously in living organisms mediated by specific enzymes. These enzymes also make it possible to replicate RNA on a template *in vitro* (i.e. in a flask or in a test tube) under laboratory conditions. But can this reaction occur without enzymes?

Numerous experiments have been performed to address this question. The results suggest that without enzymes no RNA replication can take place from the monomers used as raw materials by living organisms today.[G16] However, if other types of monomers are used instead of nucleotide triphosphates (e.g. imidazolides), RNA synthesis directed by template molecules can be carried out, although it is rather restricted. On the one hand, this is a promising result, since experiments have shown that although nucleotide triphosphates cannot be produced under the conditions of primordial Earth, imidazolides can be formed. However, on the other hand, it is not promising because in these syntheses of RNA only a few monomers combine to form a short strand, rather than the hundred or more needed to polymerize a strand long enough to code for an enzyme.

However, one factor has not been considered in these experiments. When the new RNA strand has been synthesized on the template in the experimental solution, the reaction stops because the double-stranded RNA cannot act as a template molecule until the two strands are separated. Short strands of RNA, consisting of six or seven monomers, can be separated from the template surface by thermal movement. Therefore they can continue to be synthesized without the need for a separation mechanism, but longer strands cannot. Periodically changing conditions where conditions suitable for synthesis are followed by those suitable for strand separation, are needed for the synthesis of long strands.[S14] As we have seen, in chemotons these conditions are also realized without enzymes through a periodic change of the monomer concentration. Thus there are grounds for hope that our choice of an RNA replicating system functioning without any enzymes as the information subsystem of our particular chemoton of primordial Earth is correct.

The third subsystem, the membrane, is much more troublesome. Although many experiments have shown that membrane spheres with two-dimensional liquid properties form spontaneously and in large amounts under a variety of conditions which could have existed in primordial Earth, no detailed chemical study has been performed for any of them (except for the thermal-proteinoid membranes which are not dealt with here). We do not know which compounds were used in the construction of membranes. However, one important fact that is known is that formaldehyde must be one the starting materials and must have a stoichiometric connection with the formose reaction. Apart from this, the elementary reactions leading to the formation of prebiotic membranes and their interrelations in networks are totally unknown.

Thus we can establish that we know something about the design of the metabolic network of a particular prebiotic chemoton, but it is far from sufficient. We know the abstract connection system, which needs to be completed with real chemical compounds. Two of its three subsystems could be completed with detailed chemical reactions, but we know nothing about the third subsystem except that it exists and can be reproduced experimentally. We also know of several synthetic side-systems which are related to the system, for example the formation of purine and pyrimidine bases, several other reactions, and that fact that the synthesis of RNA probably occurs through the imidazol derivatives of monomers.

This is not sufficient to fill in all the details of the abstract chemoton model with actual chemical events which could have taken place under the conditions of primordial Earth. However, enough information is available to attempt to fill in the absent details on a theoretical basis with the help of the abstract chemoton network, i.e. to establish the reactions which should be coupled to known chemical processes.

Máté Hidvégi, a student at the Technical University, Budapest, attempted to carry out this theoretical work as his masters thesis in 1981. He successfully developed a network consisting of about a hundred reaction steps, thus obtaining the complete metabolic network of a particular chemoton. About two-thirds of the reaction steps are prebiotic processes known from the literature, and the remainder are hypothetical reaction steps whose prebiotic existence has not been proved but which do not contravene the laws of chemistry. All compounds in the network are doubled on division, and the prebiotic chemoton consumes formaldehyde, hydrogen cyanide, and ammonia, and emits urea and carbon dioxide as metabolic products.

We do not claim that primordial chemotons were exactly like this, or that this system is the only possible chemoton. On the contrary, we believe that the same abstract system can be realized in many different chemical ways, and that this particular example is only one of the variants which are theoretically possible. However, it is indisputable that this work has proved that the chemoton model can be used to design program-controlled proliferating fluid automata which form spontaneously and show properties characteristic of living systems in an exact and concrete manner if the necessary data are known, just as an engineer designs his machines and instruments.

by negatively charged groups. Others, such as Orgel (2000), doubt that a single surface could have catalysed so many different reactions. They argue that *template molecules with enzymatic capabilities* should have come first, which could then have built up metabolism around them, presumably on some mineral surface. We do not know which view is correct. According to the first view, the chemoton would have been a culmination of a long phase of pyrite-bound surface metabolism; according to the second, chemotons would have possessed enzymatically active templates from the beginning. In the latter case, contrary to Gánti's belief, a non-enzymatic chemoton would be impossible. But even in this case the *regulated* system would have a chemoton-like organization, with an appropriate superimposed enzymatic system.

When this work had been completed, Eörs Szathmáry, a student at the L. Eötvös University, undertook the calculation of the chemical events of this prebiotic chemoton network using cycle stoichiometry. It was a very long calculation—a particular strategy had to be developed and some difficulties were overcome ingeniously. This calculation has proved, on the one hand, that the metabolic network is stoichiometrically correct and, on the other hand, that cycle stoichiometry is a suitable tool for the calculation of such complicated self-reproducing metabolic networks. The overall equation obtained as the end result prescribes accurately which intermediates, and how many of them, should be contained in a prebiotic chemoton with such a network in order for it to be able to operate and reproduce itself, how many formaldehyde, hydrogen cyanide, ammonia, and water molecules have to be consumed, and how many urea and carbon dioxide molecules are produced as waste.

Hence we see that if, rather than starting from the assumption that the 'texts' were formed first followed by the formation of enzymes according to the instructions written in the text, we start from the assumption that living systems are fluid machines, our view of biogenesis is changed from a non-recurring miracle to an inevitable process directed by chemical laws which can be followed by the exact methods of science.[S15] However, genetic texts, as well as the enzymes determined by these texts, are an existing reality in living systems. Our argument would not be complete if we could not provide an explanation for their formation. However, this explanation should be based on a limited series of events, on the evolution of prebiotic chemotons, and not on mere chance or statistical probabilities.

2.7   The birth of primordial texts

We have now reached a point where we can attempt to prove that hereditary properties can be divided into two categories: 'changeable' hereditary properties which form the basis of evolution, diversity, and the incredible ability for adjustment which is shown by the living world, and properties which are immutable, or change to only a very limited extent, which represent life itself—its continuity and unity despite its variability. Changeable properties are studied by modern genetics—they are written in genes (coded by a 'letter sequence') and they form the genetic program. The immutable properties are carried in the mode of organization of the construction—they are not written in the monomer sequence of macromolecules and thus they are excluded from the field of modern genetics.[G17] We shall examine the chemoton model for a clearer simpler example. Let us start with the study of changeable hereditary properties.

If we have a particular chemoton model (e.g. the prebiotic chemoton model described in the previous section), we can readily calculate how many membrane-forming molecules are required to form the membrane of a chemoton sphere of a given size. For example, if we assume that the

[S15]A further point should be added to the above considerations. Gánti (1989), in the second volume of his monograph (which has not yet been published in English), points out that the conditions under which most 'chemical evolution' experiments have been conducted are grossly unrealistic. One crucial item is the pressure of the primordial atmosphere. If we assume that the carbon from all carbonaceous minerals originating from fossils was once in the form of carbon dioxide, we obtain an estimate of 200–300 bar for the primordial atmospheric pressure. This result agrees with the fact that the carbon dioxide atmosphere of Venus (a planet smaller than Earth) has a surface pressure of 100 bar. Experiments should be conducted under such conditions.

[G17]Present-day epigeneticists consider whether some changeable hereditary properties may lie in the mode of organization of the construction rather than in the monomer sequence of macromolecules or, put differently, in macromolecules as well as the sequences of nucleotides which make up the genes. Niche constructionists consider the mode of organization of the environment in explaining the changeable properties of organisms (Odling-Smee *et al.* 1996).

prebiotic chemoton is a sphere of diameter 1 μm (the microspheres produced in prebiotic experiments are of this order of magnitude), we calculate that 10 million membrane-forming molecules are needed to covering its surface.

In order for the given chemoton to synthesize 10 million of these molecules, 10 million nucleotides must polymerize to RNA, according to the organization of the system, since the molecules (condensation products) which build the membrane-forming molecules by reacting with the membrane precursor molecules are liberated during the coupling of the individual nucleotides. However, since the synthesis of RNA occurs on the surface of a template molecule (template polymerization), only as many nucleotides can combine as are present in the template. Thus the RNA molecules of the given chemoton should be constructed of 10 million monomers (nucleotides).

The average length of an individual prebiotic RNA strand can be taken as consisting of about 100 nucleotides. Thus the chemoton should contain 100 000 simple strands of RNA consisting, on average, of 100 nucleotides, on each of which a new complementary strand is being built. Thus, immediately before division 100 000 double-stranded RNA molecules are present in the given chemoton, which are halved statistically at division (e.g. each chemoton contains about 50 000 double-stranded RNA molecules).

However, halving is only statistical, which means that the number of RNA molecules in each of the descendants is not identical; it is quite likely that of them contains only 49 900 molecules and another contains only 50 100 molecules, i.e. after separation of the double strands there will be 99 800 in one chemoton and 100 200 in the other. Obviously, the total number of membrane molecules forming in the cycle will be the same as one of the new chemotons will produces more and the other less descendants. The distribution of RNA is not accurately half and half either; thus there is a size distribution depending on the amount of the RNA molecules, i.e. the first form of genetic diversity—the genetic diversity of populations.

The total number of nucleotides in the RNA determines hereditarily the geometric dimensions of a given chemoton. As this number may vary both for the reasons discussed above and as a result of other different mechanisms, RNA carries variable hereditary properties for the whole of the system. However, the construction of RNA determines not only the number of membrane-forming molecules in the system, but also the course of the other chemical processes. Thus the RNA stock determines many other hereditary properties of the given chemoton in a variable way via the number of nucleotides and the ratio of its four nucleotides (A, G, C, U) to each other (the latter phenomenon cannot be dealt with here in detail). Nevertheless, it should be emphasized that, in these constructions, these properties are not determined by the sequence of nucleotides but by their number, and hereditary control functions stoichiometrically, i.e. through the amounts of synthesized compounds. However, the letter sequence of RNA molecules is also replicated, i.e. sequences are also inherited.

As we have already mentioned, the primitive hereditary change demonstrated leads to the appearance of genetic diversity. At this level, this means that the sizes and amounts of the internal materials of a chemoton population are not identical in the individuals making up the population and that these variations are hereditary. Naturally, under given external conditions, the operations of individuals with different internal parameters (their reactions rates) are different, and thus their proliferation rate is also different. Thus the chemoton populations can become optimized hereditarily to given conditions through a series of generations. Similarly, this also allows hereditary adaptation of the population to changed conditions. Of course, this optimization and adaptation mechanism is restricted to a narrow range of environmental variations, primarily the osmotic conditions and the ratio of the various nutrients to each other, in such a way that the hereditary properties are not yet determined by the sequence of 'letters', but only by their amounts and ratio in the 'genetic text'.

All this enables us to understand better the role of the immutable properties. We noted the division of simple proliferating microspheres involves inheritance, as two identical microspheres are formed, which are also identical with the original microsphere, but they do not contain a genetic program stored in a 'coded' form. Because of the lack of a program, this inheritance does not change. Hereditary properties are carried by the whole of the system through its construction from self-reproducing subsystems.

In fact, a chemoton is simply proliferating microsphere into which a template polymerization subsystem (which also has self-reproducing properties) is incorporated. However, whereas an autocatalytic cycle always produces an identical cyclic process, in template polymerization defects may occur during replication, which result in a new the strand being different from that on the surface on which it is produced. Hence template polymerization is a self-reproducing process which allows variation in the length, composition, and monomer sequence of the polymer in the descendants.

The raw material for template polymerization is produced by the autocatalytic reaction network of the chemoton. If the number, composition, or length of the template polymers changes, their requirements for raw materials also changes and thus they have an influence on the quantitative relationships of the autocatalytic reaction network as well as on the amount of membrane building blocks. Thus changes in the template molecules in these chemotons may also cause hereditary changes in the other parts of the chemotons. A change in the monomer sequence of templates (i.e. in the base sequence of RNA) does not change the raw material requirements for the template polymerization process, if the composition does not change. Thus the sequence does not play any role in these constructions. Therefore a change in the template can change the size of the chemoton, shift the ratios of operation of the various chemical pathways, or stop existing chemical reactions. Namely, in a given construction, their presence or absence is restricted only by chemical possibilities.

The situation is different in the present living world, where each reaction step takes place in the presence of enzymes. Enzymes are capable of accelerating the rate of possible chemical reactions by a factors of a million or tens of millions. If the reaction network consists only of enzymatic reactions, the operation of the network, within chemical constraints, is completely determined by the presence or absence of enzymes, since a non-enzymatic reaction, which is a million times less efficient than an enzymatic reaction, is obviously negligible in the system as a whole. Thus the present living world is regulated by the presence or absence of enzymes, not only by the productivity of individual reaction pathways but also by the quality of chemical reactions taking place within the constraints of the laws of chemistry. As we have already indicated, this regulation occurs via the amino acid sequence of enzymes which, in turn, is determined by the 'letter sequence' of the nucleic acid in the genetic material.

Here we arrive at a point where the critical question can be put: How did the primordial letter sequences come about? In other words, how did the primitive program control of our proliferating, regularly operating, and program-controlled machine change into an 'up-to-date' program control? We need not invoke miracles or blind chance any more, but should establish whether the evolution of chemotons, these program-controlled proliferating fluid machines, has led to the development of specific functional 'letter sequences' which can be used advantageously in such a system. The results of such studies by Gánti were published in 1979 and 1983, and they are given in more detail in the second volume of *Chemoton Theory*. Here, we provide a summary of the chain of thought followed in these studies.

Perhaps we should start by envisaging a prebiotic chemoton population proliferating in a primordial nutrient soup. The chemotons have a diameter of about 0.001 mm and each contains about 50 000 double-stranded RNA molecules with a length of about 100 nucleotides. What are these RNA molecules like? At first sight (of course, at a 'submicroscopic' sight) they have a double-helix structure, similar to that of the Watson–Crick model of DNA. The small difference between them is of no interest to us at present. Every strand is built from four types of nucleotides (A, G, C, U). The two strands are linked through the nucleotide pairs A–U and G–C. This means that if there is an A (adenine) some where in one of a pair of strands, there must be a U (uracil) opposite to it on the other strand, and vice versa. Similarly, if a G (guanine) is present on one strand, then there must be a C (cytosine) opposite to it on the other strand, and vice versa. For example, if on one strand the nucleotide letters are arranged in the sequence GACUGA, then the sequence on the other strand of the pair must be CUGACU.

What was the letter sequence of RNA molecules in the first chemotons? Obviously, it was completely random, without any information, similar to text typed by a blind man.[S16] It was also different for each RNA molecule, unless the 'descendants' (replicas of the same RNA) entered the same chemoton during division, but we shall ignore this

[S16]It may not have been completely random. First, the assembly of the first RNA strands could have been biased by chemistry to favour some sequences. Second, a long phase of evolution (e.g. of mineral surfaces) could have preceded the chemoton era.

possibility for the time being. However, it must be remembered that all unintelligible texts exist in two copies—a 'positive' and a 'negative'— since the double-stranded structure determines the sequence of both strands. If the double strand separates and a new strand is synthesized on each of the original pair, then the negative sequence is built on the positive strand and vice versa. Thus a CUGACU sequence is synthesized on the GACUGA strand, and a GACUGA sequence is synthesized on the CUGACU strand. Hence each sequence could carry the same information, if these sequences had any sense which, as yet, they do not.

Let us now see what happens to the RNA molecules in a chemoton when the concentration of nucleotides reaches the critical value for separation of the double strands. The double-stranded structure is quite rigid; it can be bent only on a longer part of the structure. However, the separated strands are very flexible, like a chain, and they can twist, wriggle, or bend. Thermal motion causes them to twist a million times per second. If they could be observed under a super-microscope, they would look like cheese mites, with their motion accelerated a million times. However, if the movement could be slowed down by a factor of a million, we would see that as the RNA molecules twist and wriggle, nucleotides swarming around them are coupled to this strand and to each other by choosing the appropriate partner. This continues until the last nucleotide of the sample strand finds its partner, thus completing the synthesis of the new strand, forming a double-stranded structure and doubling the (still senseless) information, and become as rigid as a corrugated metal tube.

If we had a microscope which would enable us to observe this series of events at a magnification of a million (which is not impossible due to the development of electron microscopes), we would be fascinated by this sight. However, if we observed it for a sufficiently long time, we would see that not all RNA molecules behave in this way. The synthesis of the new RNA strand is sometimes completed in another way (this may not be a rare occurrence). The nucleotides forming RNA have two strong valences capable of chemical binding rather like the two arms of a human. We can hook onto two people simultaneously, and each of our neighbours can hook onto another person, etc. Nucleotides can do the same—each nucleotide can hook onto two other nucleotides, which can hook onto two more nucleotides, etc.—and this is how the RNA chain is built. Both 'arms' of each nucleotide are occupied, except for the two nucleotides at the ends of the chain.

Let us now recall the picture seen under the super-microscope. The RNA molecule strands, i.e. the nucleotide chains, are twisting and wriggling to and fro. The free nucleotides choose a pair on the one end of the strand, and then they hook onto the nucleotide which is already bound. Then another suitable nucleotide comes along and hooks on, and so on. Simultaneously, the free part of the template strand wriggles and twists, and during one of these wriggles the end of the sample strand and the last nucleotide incorporated, which has a free arm, touch each other. The two free arms interlace and thus the synthesis stops because there are no free arms. Accurately speaking, there are no new and old strands here, since the old and new strands are coupled together into a single strand (Fig. 2.6). There has been a silly misbirth—a strand whose two ends form a double-helix structure by themselves; a strand whose front and back ends carry the same information, only through complementary sequences, i.e. if GACUGA appears at the front end then CUGACU is developed at the end. Is this information still unintelligible? Not at all. The 'letter sequences' at the two ends of the strand carry the information that the two ends of the molecule should form a double helix so that the middle of the chain forms a single-stranded loop. This means that the 'letter sequence' determines the space structure of this polymer. And what did we say about enzymes initially? We stated that their amino acid sequence determines their space structure which in turn determines their function, and we made first step. We have found a spontaneous mechanism that is suitable for producing RNA molecules with a structure determined by a sequence. Moreover, they are produced in a series as the beginning and the end of a copy built on a strand are also capable of building a pair, as well as its copy. This structure, which is called loop RNA, is particularly important in present-day living organisms.

There is another step left. We already have spatial structures of macromolecules determined by a sequence, but these spatial structures do not have any function in our chemoton. This does not matter. Until now, we had sequences without sense. Now we have sequences with sense, and we achieved this simply by observing what is going on in the chemoton under a super-microscope. Let us observe further: Perhaps the solution will be self-evident.

In a single chemoton, before a single division, new strands are synthesized on 100 000 RNA strands. Thus it can be assumed that several loop RNAs are formed during each division, which then appear in the descendants. At the next division new loop RNAs form, and obviously the older ones are also replicated, and this process continues. It can easily be seen that, once chemotons have appeared, their population goes

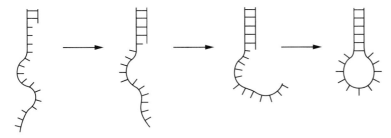

Fig. 2.6 Abiotic loop RNA formation. The template RNA which swirls to and fro due to thermal motion is made rigid by building a new strand on it. However, the free end of the template strand can be coupled to a partially formed new strand; thus an RNA loop is formed. The copies of such loops have also a loop structure. The result of this process is that the majority of the molecules formed in abiogenic RNA synthesis have a loop structure.

through the evolution process very rapidly. Thus there are a large number of loop RNAs in chemotons.

Let us now observe such a chemoton under our super-microscope. We would see that thermal currents keep the loop RNAs in continuous motion such that they collide with either the different molecules of the metabolic network or other RNA molecules.

First, we consider collision with another RNA molecule. If part of the double-helix structure collides with another double helix, nothing happens and the two molecules rebound since no free valences or free secondary binding sites are available. If two loops collide, there is short-lived binding between them because part of the loop is a single-stranded structure in which the weak binding sites of nucleotides which enable them to form pairs are free and can form weak bonds with other appropriate atom groups. However, these bonds are quickly broken by thermal motion. Rarely, although with a well-defined probability, loop RNAs can also collide such that their ends come into contact. In this case, since there are free valences at the ends of the RNA strands, the two RNA loops can combine.

Such a series of couplings is shown in Fig. 2.7. In the first step, two loop RNAs combine, and in the next step a third loop is coupled to the strand containing two loops. If we the structure obtained is flattened into a plane, we obtain a clover-leaf shape, which is the same as that of the transfer RNAs, which play a central role in the metabolism of present-day living organisms. They are the maids-of-all-work of protein synthesis. They are charged by the cognate amino acid and they recognize and read the coded messages from DNA in order to couple the appropriate amino acids prescribed by the genetic message to the protein chain being built.

Since no protein synthesis takes place in our present chemoton, the randomly formed clover-leaf RNA has no function as yet. However, it already has a very complicated spatial structure, since in the chemoton it is not present in its flattened planar form, but in a curled-up spherical form. The spatial structure of this complex formation is strictly defined by the 'letter sequence' of the strand, and all its copies will have the same spatial structure.

The letter sequence determines not only the spatial structure, but also which of the secondary binding sites in the loop are on the surface of the 'ball' and what position they take there. This is also an important property of enzymes: their 'letter sequence' (amino acid sequence) determines the spatial structure of the protein molecule, which secondary binding sites are located on their surface, and where they are located. Thus we are already quite close to the development of a sequence-dependent function.

Let us now return to our super-microscope and observe what happens if our RNA loop collides with a small molecule of the metabolic network rather than with another loop (the former is a more frequent event). The small molecules participating in metabolism also have secondary binding sites through which they can temporarily combine with other small molecules or with macromolecules (enzymes, loop RNA). The operation

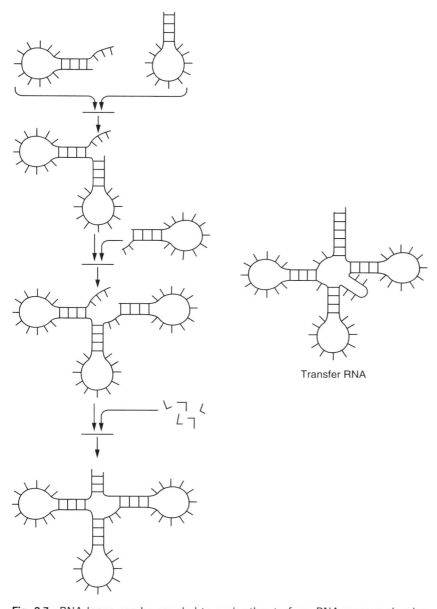

Transfer RNA

Fig. 2.7 RNA loops can be coupled to each other to form RNA macromolecules with a very complex spatial structure, which is determined by the nucleotide sequence in the RNA. When these replicate, both the sequence and the complex spatial structure are copied. Such spatial structures have many specific functions. The right-hand side of the figure shows the clover-leaf structure of transfer RNA (which plays a key role in protein synthesis). Development of such a structure by abiogenic coupling of three RNA loops is shown on the left-hand side.

of enzymes is based on the phenomenon that, in a suitable position, they 'push' small molecules which are temporarily coupled to each other due to their particular spatial structure into a position such that they can form strong chemical bonds or, in contrast, they can hold a large molecule temporarily, thus distorting its electronic structure so that it can transform or split into two smaller molecules.

For simplicity, we shall consider the latter case first. Let us take a large molecule with two temporary binding sites at both ends that is to be split into two smaller molecules through reactions with the metabolic network. As this molecule and its neighbours are agitated by thermal motion, one of its ends becomes temporarily coupled to other molecules. These other molecules could, of course, be loop RNAs. If loop RNAs of the appropriate lengths are eventually coupled to both its ends, these can be bound by chemical bonds since the small molecule brings them together. At this point, the first enzyme is formed. This enzyme is not a true protein, but RNA, but it is capable of functioning in the same way as the appropriate protein enzyme.

Let us consider further what happens after the coupling of the loop RNAs. Owing to changes in electronic structure, the molecule bringing them together splits and the parts move away from the RNA via thermal motion. However, the structure consisting of two coupled loop RNAs remains, with its secondary binding sites located spatially such that they are capable of trapping molecules identical with those bringing them together (which then split and separate from them). Thus they become splitting enzymes, and the molecule bringing them together becomes their substrate. The whole process is illustrated schematically in Fig. 2.8. Enzymes capable of coupling two small molecules into one larger molecule (Fig. 2.9) or transforming a molecule are arranged similarly.

Thus we have proved that enzymes need not have been formed by some miraculous chance; any of the molecules involved in metabolism could assemble its own enzyme in a chemoton crowded with loop RNA.[S17] If an enzyme is assembled to accelerate a reaction step of metabolism in a chemoton, then our computer calculations have shown that this chemoton develops a little faster and divides a little sooner than those without enzymes; thus its descendants multiply faster. Then another enzyme will be assembled in one of its descendants, a third enzyme will appear in its descendants, and so on. It seems that chemotons could reach the point where each of their metabolic steps was catalysed by a separate enzyme during a very fast evolution process. However, it must be remembered that they are not protein enzymes, but enzyme RNAs (ribozymes).

Here we must again pay tribute to Francis Crick. He was the first to propose, more than three decades ago, that RNA molecules might have served as enzymes during biogenesis. However, he did not reach this conclusion via the same chain of thought as presented here. The present author inferred this on the basis of chemoton theory a decade later. However, the scientific establishment rejected these ideas, claiming that there was no example of RNA playing the role of an enzyme. One more decade has passed since then. A decade ago, Cech and Altman made the

[S17]This theoretical process must be complemented by selection acting between chemotons. Chemotons equipped with a better RNA set will reproduce faster. Mutations in RNAs will be selected for if a required reaction is catalysed more efficiently. Selection will act against chemotons where RNA structures happened to have assembled catalysing an *unwanted, detrimental* reaction. There is nothing *within* the chemoton that could act against such a malevolent enzyme, but selection *between* chemotons can be effective.

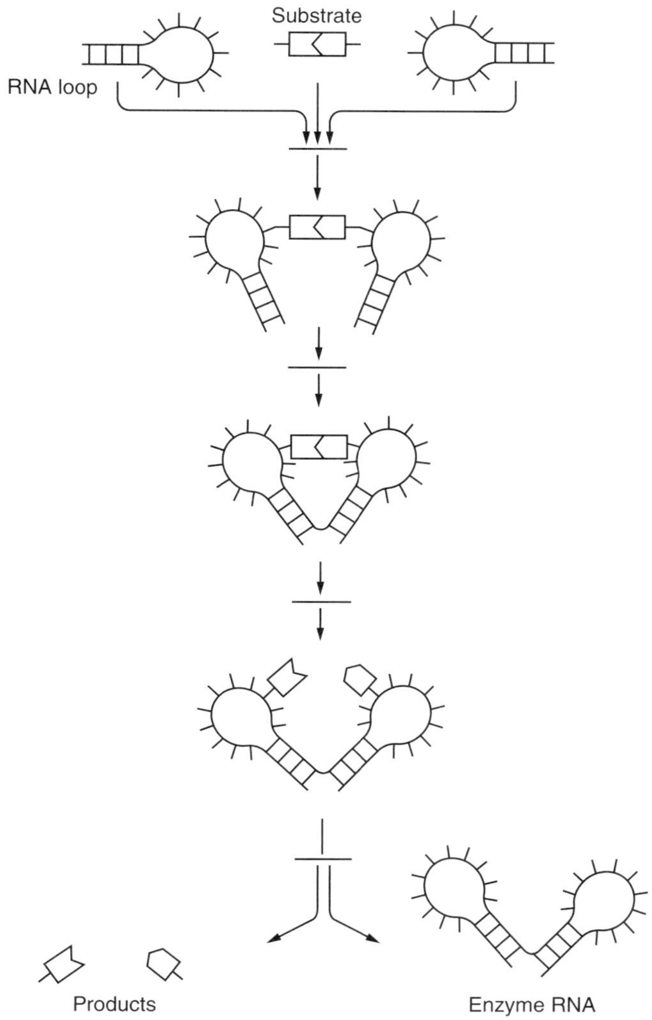

Fig. 2.8 The coupling of loops need not be based on pure chance. The loop contains many active atomic groups suitable for secondary binding, which can make a temporary bond with the molecules participating in metabolism. This enables these compounds to assemble their own enzymes from RNA loops. The figure shows schematically how a molecule with two binding sites is capable of assembling an enzyme RNA, which is able to split it specifically.

discovery that enzyme RNAs are operating in the living world today, for which they were awarded the Nobel Prize for chemistry in 1989.

2.8 We crossed the 'finish line'

We have arrived at our goal. We have shown how self-reproducing fluid machines in primordial water and then dividing machines in space have

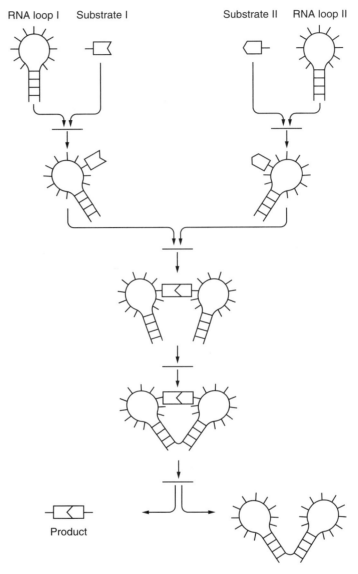

Fig. 2.9 Synthesizing enzymes could also be assembled from RNA loops in an abiogenic way. A diagram of such a process is shown.

been developed, how primitive program control and then sequences have come about, how the randomly formed sequences have gained sense through loop structures, and how function has appeared from these. In other words, we have shown how enzymes and genes have been created.

The attentive reader might say that so far we have not found the solution for the development of genes. However, we have found it. We have the solution, but we have not yet discussed it. Let us consider what is happening in the replication of an enzyme RNA. First, the double-stranded structure is broken and the RNA becomes single-stranded. At the same time its special spatial structure is also decomposed, and the uncoiled strand cannot function as an enzyme but now operates as a template for the synthesis of a new RNA strand. The letter sequence of this new strand is obviously determined by the template according to the rules of pairing, i.e. where there is an A in the template, there is a U in the new strand, where there is a G, there is a C, etc. Hence the sequence of the new strand is not identical to, but complementary with, the template strand.

Let us call the strand with the enzyme function positive and the complementary sequence negative. Obviously, the spatial structure of the negative strand is the same as that of the positive strand, since where the positive strand can be helicalized with itself, so can the negative strand. However, it is also obvious that it does not have the same enzyme property as its positive counterpart, since where there is an A on the former, there is a U on the latter, and so on, i.e. the functional groups reacting with other molecules are located differently on it. Thus we have two RNA strands: a positive strand and a corresponding negative strand. The positive strand has two functions: when it is coiled it serves as an enzyme, and when it is an open single strand, it plays the role of a template for the negative strand. But what is the function of the negative strand, if it has no enzymatic role?

It is the gene, i.e. the positive strand is synthesized on it during replication and it carries the information necessary for the construction of the given enzyme (e-RNA). This is its only function, but this function is decisive in the living world. If there is a change in its sequence, the sequences, design, and operation of the enzyme determined by it also vary. This means that not only has the gene been created, but so has the mechanism of hereditary change known as mutation which also plays a central role in modern genetics. The next stage of evolution can be reached via this mutation mechanism.

The RNA replication mechanisms described here are very imperfect, i.e. the probability of faulty incorporation of nucleotides is very high. This can be expressed by saying that the mutation rate of these systems is very high. This accelerates the process of evolution, thus making possible the appearance of more perfect enzymes and their associated gene systems which operate better and are more efficient. This means that the chemotons of primordial Earth could have completed the optimization of their enzymes in a very short period of evolution, resulting in a level of evolution where every steps in their metabolism was catalysed by an enzyme (e-RNA) and every enzyme was highly adapted.

Just as a high mutation rate was an advantage[S18] at the beginning of the optimization process by enabling the appearance of newer and newer variants for selection, it became a disadvantage at the end of optimization. As the system became perfected during this process, almost every change could only impair its operation. It would have been very favourable to retain the information content of optimized genes. This necessity might have led to a process during which the sequences of genes were transcribed into DNA molecules with a more stable structure and the catalytic role was taken over by more suitable macromolecules—the proteins. RNA

[S18]Most mutations are harmful. Recurrent mutations imply a so-called mutational load, recognized by population geneticists a long time ago. The flip side of the coin is that a high mutation rate limits the genome size that can be maintained by selection, as recognized by Eigen (1971). Nobody has calculated the tolerable mutational load for a chemoton-like system.

molecules have remained, as information carriers, until the present day, at least in the terrestrial living world. The mechanism of this process is not known at present, not even schematically. However, the chemoton in which hereditary information is carried by DNA and whose metabolic processes are catalysed by enzyme proteins, is called the cell—in the most biological sense of biology.

We started with chemical processes taking place in the atmosphere of primordial Earth, and have ended with the living cell through a logical succession of steps. In this long series of events we have never needed to invoke a miracle, blind chance, or an improbable event. If there is a miracle in the genesis of life, then it is how matter is organized into ever more complex systems with ever more functions, by following the laws of nature, up to the point where the most complex individuals of this system—humans—are able to discover and understand the laws of this organization.

It is not at all certain that everything happened in exactly the way described above and that all the details are correct. However, there is enough evidence available to demonstrate that if we do not insist that proteins were formed without a machine and that hereditary programs developed without hereditary systems, and that if we do not accept modern biological theory as dogma, then a clear unambiguous answer can be given to the question of the genesis of life. This answer is based partly on the double-helix molecular structure discovered by Watson and Crick, in which the enzyme RNAs first proposed by Crick play an important role.

Finally, we must admit that our proposition, in the form given here, is not science. Rather, it is a logical argument. Exact science includes, in addition to logical reasoning, evidence, exact precise theoretical derivations, and quantitative calculations. However, the proposition presented here is underpinned an exact scientific theory with evidence, derivations, and calculations. Interested readers are referred to another book by Gánti entitled *Chemoton Theory*.

3 The unitary theory of life

Gánti, T. 1987, in Hungarian 1978

Life as a whole, from the simplest organisms to the most complicated ones—man of course included—is a long sequence of counterbalancings with the environment, a sequence which becomes eventually very complex. The time will come—even if it is still rather far away—when mathematical analysis supported by scientific analysis can put into majestic formulas of equations these counterbalancings, putting finally into an equation life itself. (Pavlov)

3.1 Exact sciences

What do we understand by the term 'exact science'? Exact sciences, such as mathematics, mechanics, the theory of electricity, thermodynamics, chemistry, etc., are characterized by the common fact that they all have specific model systems, i.e. systems which represent phenomena of the real world without disturbing factors. Thus exact sciences are distinguished by being capable of constructing the specific phenomena under investigation in a pure form.[G18] By using model systems the exact sciences can describe the phenomena under investigation in qualitative and quantitative respects, and can formulate them in mathematically.

The above definition contains two points worthy of consideration. First, we have to emphasize that *any one of the exact sciences models only one part of the real world and even this one only from a definite point of view, independently of the other phenomena.* Second, it must be understood that *it is not the real world which the exact sciences are capable of treating with an arbitrary? exactness, but their own model systems.* Real-world phenomena are only approximated by them. This approximation makes the model system of an exact science intricate, and the mathematical apparatus of the exact sciences becomes more complex, the more precisely the real world is intended to be approximated by the model system.

Some examples will make this more clear. The basis of mathematics is the series of natural numbers. Such a series does not exist in the real world, since it is based on the abstract idea of completely identical objects, which is an absurd assumption in the real world. If we take such an object—just one—and then a similar object—again, only one—these two together give two. This is the start of the series of natural numbers: one plus one is two. However in the real world, there are no two objects which are completely identical; hence the most fundamental concepts of mathematics, unity and the series of natural numbers, do not exist in the real world. Rather, they contradict it, thus remaining merely absurd abstractions. Nevertheless, these abstractions are very fruitful; without them it would be impossible to give a quantitative description of the world.

[G18]Philosophers of science describe this pure form of representation of constructed phenomena in terms of the idealization and abstraction of models from phenomena (Cartwright 1983, 1989; Giere 1988, 1997; Morgan and Morrison 1999).

Geometry is based on abstractions which are just as absurd. Its basic concept, the geometrical point, is a twofold absurdity: on the one hand, the point contains within itself a basic absurdity—its unity—since every geometrical point is a unit and, as a unit, is identical with every other geometrical point. On the other hand, it also includes the basic absurdity of geometry, namely that of being without extension; it has no extension but nevertheless is something. However, it is only this absurdity which allows spatial existence and motion to be exactly described by means of geometrical concepts, since the possibility of error is completely excluded from the explanation by the very extensionlessness of the point. The other fundamental concepts of geometry are just as absurd—for example the line, which by definition extends only in one dimension, the plane which extends in only two dimensions, or the meeting of parallel lines at infinity. Infinity itself is another useful 'absurd' abstraction.

Despite these absurdities, or more exactly because of them, geometry can be treated as a model world with an arbitrary exactness. Only in this way can the properties of a rectangle, a circle, an ellipse, a cube, a cone, a sphere, etc. be described exactly by mathematical methods.

However, it must be understood that these are mere abstractions; no lines, parallels, planes, rectangles, cubes, cones, spheres, etc. exist in the real world. By using this model system to describe the real world, geometry approximates real phenomena and events only as far as is necessary for the given problem. However, this approximation is exceedingly effective; it is sufficient to solve most of the problems of everyday life.

There are just as many contradictions in mechanics as in arithmetic and geometry. Moreover, the fundamental concept of mechanics—the mass-point—is the source of further contradictions. Since it is a unit, which is itself a contradiction, it is without extension, i.e. it contains the contradiction of geometry, and it also has mass despite being without extension. This is a further contradiction, since in the real world no matter exists which as mass without extension. But this contradiction is necessary for the model system of mechanics to represent mechanical phenomena sufficiently well to approximate real-world phenomena to a degree suitable for practical purposes.[G19] The approximation can be surprisingly precise: perhaps the best example is space technology, when a rocket reaches its target after travelling hundreds of millions of kilometres with an accuracy to a hundred metres.

The theory of electricity starts from similar abstractions since, to remain with points, point-like electric charges are fundamental to the theory.

It is not necessary for the fundamental principles of every exact scientific discipline to be reduced to such prime absurdities. The fundamentals of chemistry can be reduced to concepts readily accepted by other sciences (mathematics, physics) and thus the absurdities in its model system only appear indirectly. But at the inception of chemistry, its fundamental concepts (element, compound, atom, molecule, valence, chemical bond, etc.) were abstractions, although not absurdities. Later on it was found that these abstractions could also be found in the phenomena of other sciences, and therefore they appear to us as realities.

[G19]Philosophers of science have described models in terms of idealizations (all else, besides what is the subject of the model representation, being equal, or *ceteris paribus*) and abstractions (all else being absent, or *ceteris absentibus*). What is being discussed here is a stronger property of abstract idealized models in the exact sciences—that what they represent as possible, i.e. occurring in the model world, is something physically impossible. The claim is that these contradictions—absurdities—are *necessary* for models to represent reality approximately.

Biology is mostly a descriptive and experimental science. Exact theoretical biology must develop its own fundamental concepts and must define these with an axiomatic precision, starting from the axioms and building up a model system which forms a basis, at least in principle, for any phenomenon occurring in the living world and which can be described both quantitatively and qualitatively by mathematical methods.[G20] In fact, these requirements are nothing less than the birth of theoretical biology, waited for in vain by the science of life for so many decades.

Undoubtedly, books with the proud title 'theoretical biology' have appeared over the decades and the *Journal of Theoretical Biology* is a well-respected international publication. However, in the proper sense of the phrase, we cannot yet speak of the birth of theoretical biology. Disciplines bearing this name are the application to biology of results obtained from other exact sciences (mostly from mathematics, physics and cybernetics) with the aim of providing, as far as possible, a quantitative description of the phenomena. However, this is far from being an autonomous theoretical biology.[S19]

Theoretical biology in the proper sense must be grounded in a model system based upon abstract models of the simplest living systems. Models of more complex biological phenomena must be deducible from this system by means of pure logic, and they must be expressible by mathematical formulae.[S20,G21]

As yet, fundamental models of this kind have only been developed in biology for partial problems. For example, a simple key–lock model for immune reactions and enzyme–substrate specificity served quite well for a long time although it could never be expressed mathematically and later on was found to be oversimplified and mistaken. The DNA model developed by James Watson and Francis Crick was far more fruitful. A whole new discipline—molecular biology—was generated by it, building an extremely successful abstract model upon the basic unit of genetics—the gene.[G22] This abstract model enabled storage and transfer of hereditary properties to be interpreted logically. Moreover, another equally useful model—the Lwoff–Jacob–Monod model of genetic regulation was built on the Watson–Crick DNA model. These two models changed the fundamental view of biology in less than a decade. The new discipline of molecular biology opened up entirely unsuspected perspectives on the research of life.

However, the DNA model and the model of cellular regulation are quite unable to express quantitative properties adequately.[S21] Apart from their specificity, it is mainly because of this failure that theoretical biology needs some other model system which can represent both qualitative and quantitative aspects of phenomena. It seems that the theory to be introduced below—the chemoton theory—succeeds in this task.

As with every exact model system, chemoton theory has its scope of applicability—it is restricted to the level of biological individuals. Life is structured according to an organizational hierarchy. Living beings as living entities can be prokaryotes, eukaryotes, or multicellular

[G20]Philosophers holding to the 'semantic view of theories' (van Fraassen 1980; Lloyd 1988) distinguish between the presentation of a theory by means of axioms and by means of models. The basis of this distinction was Tarski's demonstration in the 1930s that the models of a theory satisfy the theorems that follow deductively from its axioms, so that a theory can be specified by means of either models or axioms. It is a matter of controversy among philosophers of science whether this formal approach to models describes the kinds of things scientists talk about when they refer to models.

[S19]This statement is rather extreme. Population genetics and theoretical ecology are established parts of theoretical biology.

[S20]A unified theoretical biology may be just out of reach for a long time (perhaps for ever); just think of the difficulty in trying to deduce the laws of thermodynamics from statistical physics, or the lack of an accepted unified theory of physical forces. Biology deals with more complex systems than those dealt with by physics.

[G21]One can interpret this type of call for theory reductionism (that the goal of science is to formulate general theories from which the others are derivable) as a methodological strategy—that an effective way to make theoretical progress is to pursue the goal of deducibility even if it is never achieved. Wimsatt (1987), for example, argues that the engineer's use of such a strategy is to learn from the inevitable failures of reduction by systematically studying how the deduction breaks down.

[G22]Of course, much more was involved in the development of the discipline of molecular biology than the Watson–Crick model of DNA. Historians have pointed to a variety of contributing biological traditions (Olby 1974; Judson 1979) and institutions (Abir-Am 1985; Kay 1993).

[S21]There is an emerging theory of genetic circuits, which are able to generate testable predictions (Szathmáry 2001b, and references cited therein).

[G23]Many philosophers of units of evolution also consider higher-level entities such as families, demes, populations, and species to be potential individuals (Hull 1988).

organisms.[G23] Chemoton theory tries to treat these individual units theoretically, mainly at the prokaryote level. However, it does not pretend to deal with phenomena ruled by statistical laws. Thus different methods are required for investigating problems of population genetics or describing phenomena of biocenoses. Notable results have recently been obtained in these domains, and fruitful mathematical models have been developed using theoretical biology. However, we do not deal with these here; our investigations are concerned with the laws governing the function of individual organisms which are presented here on the basis of the chemoton theory together with some results derived from it over recent years. Thus in the following we shall introduce the chemoton theory in its present state of development. However, it must be emphasized that we cannot claim that the theory is in a complete and finished state. The variability of life is truly inexhaustible. Surveying the whole, taking into account every phenomenon, and formulating the appropriate definitions so that they will never contradict any biological phenomenon and at the same time will be relevant for the whole of life comprises a task which cannot be fulfilled in one or even a few steps. This aim can only be approached through continuous revision and correction. Thus it is probable that the present formulations will be further revised in the future, although the fundamentals may not change.

3.2   The units of life

Every exact science is based on 'units' of some type. By 'unit' we do not mean the units of the metric system for various disciplines, but those basic simplest entities which still retain the characteristic properties investigated by the discipline in question. For example, in this sense the units of geometry and mechanics are not the metre and the gram, but rather the point and the mass-point respectively. However, let us consider this in more detail, using chemistry as an example.

If we halve a glass of water, both parts have the properties of water. If we continue the process, producing a quarter, an eighth, a sixteenth, a thirty-second, etc. of the glass of water, the fractions obtained will always have the properties of water. However, this procedure cannot be continued infinitely. Eventually we will reach a certain minimal unit which still has the general properties of water, but when this is halved the properties of the resulting parts are no longer characteristic of water. In this way chemists arrived at the idea of the molecule and established the real purpose of chemistry—the investigation of qualitative and quantitative properties which appear when a molecule is produced from other smaller components or, conversely, which disappear if a molecule disintegrates into such components. Therefore the molecule is a basic theoretical unit of chemistry in the sense of a principle.

Crystallography provides a similar example. The crystals of common salt (sodium chloride) exhibit characteristic symmetries: the angles between their lateral faces and between their edges are constant, and although their optical properties depend on direction, they are the same from a given direction in every sodium chloride crystal. If a sodium chloride crystal is broken into pieces, each of the small crystals obtained has the characteristic crystallographic properties. However, this process cannot be continued indefinitely. Eventually we arrive (in thought, although never in practice) at a minimal piece of crystal which still shows all the characteristic properties of common salt but these disappear if this smallest piece, the so-called elementary cell, is broken again. These elementary cells are the basic units of crystallography.

What are the basic units of biology?[G24] The answer seems to be quite simple: the minimal units which are still living, but which, if disintegrated, lose the properties characteristic of life. This simple principle, if applied to the practical questions of the living world, leads to many contradictions. For instance, if we cut the throat of a chicken, the bird ceases to live. During the Middle Ages beheading was a very common punishment and the convict immediately ceased to live. From this it follows that the chicken or the man would be the last units of life. However, if only a foot or a wing of the chicken is cut off, or a man loses a foot or a hand in an accident, the properties characteristic of life are not lost—the chicken or the man do not cease to live.

Contradictions of the opposite nature may also occur. For example, earthworms and fresh-water polyps which have been cut into pieces, 'regenerate' themselves: each piece grows into a complete animal. Thus we must ask the questions: Is half an earthworm also alive? Is a tenth of a polyp alive? Or think of the example of the experiment performed by Albert Szent-Györgyi: a heart removed from a frog beats and functions for several hours, provided that a solution of an appropriate composition continuously flows through it. Is the heart prepared from the frog alive? Common sense suggests that it is, although in this sense it is called 'surviving'. However, if this heart is alive, what is really living—the frog or its individual components? If we cut the frog heart into pieces, the separate pieces of muscle still function—therefore they live.

Tissue culture is an indispensable tool of virus research and vaccine production. Tissue cultures are obtained by growing pieces cut from an animal or a plant on an artificial nutrient solution. In this culture the individual cells of the tissue are nourished, function, and multiply further. Tissue cultures can be prepared from pieces of tissue cut from the animal after its death, and the cells obtained can be maintained and propagated further for a long time. What is living—the animal, its tissues, or its cells?

For millennia the criterion of human life was thought to be the heartbeat. Until the heart stopped beating, the person was considered to be alive; when the heart ceased to beat, he died. Thus the functioning of the heart was not a criterion of life itself, but of human life in particular. But heart transplantations have recently become a common occurrence, completely destroying this view. The heart of a living man is removed during the operation and is exchanged for the 'living' heart of a dead man—and the patient continues living.

[G24]Huxley (1912) considered the nature of biological units in terms of the halving of an organism, a tradition continued by those biologists working on 'aclonal' (single-celled) and colonial organisms (Jackson *et al.* 1985).

But could a murderer defend himself before a court by reasoning that he did not deprive his victim of 'life', since the victim's cells could be made to survive in a tissue culture? Could he pretend that his victim is 'living', because the victim's heart or kidney was transplanted and is continuing to live in another person? And what if 'cloning' is successfully achieved in humans just as it is now possible in plants?

Obviously, our point of view fails somewhere. We have lost the definition of the limits of life. A hundred years ago these ideas were uniform: the lion died if it was shot, the tree perished if it withered. However, now that viruses are crystallized and even synthesized, when cells removed from dead organisms are maintained for decades in tissue cultures, when a complete organism can be developed from a single somatic cell, and when a man's 'life-giving' organ can be taken out and exchanged for the 'living' organ of a dead man the concept of 'living' clearly needs further investigation.

In order to give an act definition of the ultimate living units—the fundamental units of biology—we need to clarify some further concepts. These will be introduced in the next two chapters. Only then will we be able to give an exact definition of a living system.[G25] Meanwhile, as a first approach, we confine ourselves to the units of life and do not strive to be completely exact.

Let us start from the contradiction that after the death of a multicellular organism its parts can behave as if they were living.[G26] For instance, a frog is alive whilst it displays the properties characteristic of life: nutrition, movement, respiration, and sensitivity. Obviously all these phenomena can be abolished by destroying the spinal cord of the frog. Therefore this procedure abolishes the functioning of this particular living unit.

However, if we transplant cells from the frog's intestinal mucosa into a tissue culture, under appropriate circumstances they will continue to metabolize, retain their sensitivity, and also multiply. Although we have abolished the functioning of a particular living system, i.e. we have transformed a living unit into a non-living one, we have encountered many new living units at this lower level of organization, i.e. namely at the cell level.

Cells show a rather complicated structure under an optical microscope. They contain minute particles, such as mitochondria, with a further inner organization visible only under the electron microscope. The mitochondria multiply within the cell independently of cell reproduction and have considerable genetic self-determination.

If the cells are mechanically disrupted—if they are killed—they cease to be living units. However, if the mitochondria of these disrupted cells are collected by a suitable method and placed in an appropriate environment, independent of the cell, they continue to function, performing metabolic activity and oxidative processes, i.e. the same processes which were their basic tasks within the cells.

Are these isolated mitochondria alive? The question is not easy to answer. Nobody has yet succeeded in propagating isolated mitochondria.

[G25]There are two ways of understanding the role of 'exact definition' here. One is to assume that an exact definition specifies the conditions necessary and sufficient for something to be alive or, in other words, to give an analysis of the concept of a living system or being alive. The other way of understanding exact definitions, in line with the discussion of models in the previous section, is that they serve to introduce theoretical models—abstract representations to which nature is only approximated in some respects and to some specifiable degree (see Giere (1988, 1997) for further philosophical analysis of the relationship between models and definitions).

[G26]A good non-technical description of the processes of multicellular death is given by Clark (1996).

However, chloroplasts, the organelles performing photosynthesis in plant cells, have already been observed to divide extracellularly in a suitable medium, although only once.[S22]

Thus, life has units on at least three distinct levels of organization: on the level of certain cell organelles, on the level of cells, and on the level of multicellular organisms constructed from cells. All three levels have their own distinct definitions of life and the cessation of life: life at one of the three levels does not directly abolish the life of the units of the other two levels. Our reasoning has progressed down this hierarchy of levels from the complex to the simple, but it will also be useful to reverse this process.

The universe of unicellular organisms used to be classified into two large groups known as 'prokaryotes' and 'eukaryotes' with the former not having such a well-structured and separated nucleus as the latter. The prokaryotes are the simplest, and supposedly the most ancient, living beings; they include bacteria and cocci, blue algae, and spirochetes. Prokaryotes have neither mitochondria nor chloroplasts.

It is now accepted that thousands of millions of years ago, some amoeba-like ancestral cells, which were only capable of fermenting readily available organic material, absorbed oxygen-breathing prokaryotic cells which retained their relative integrity and self-sufficiency and changed into mitochondria. Animal cells could also have originated in this way. Similarly, if fermenting ancestral cells also absorbed, in addition to the oxygen-breathing prokaryotes (the later mitochondria), some photosynthesizing blue-green algae as the primeval chloroplasts, plant cells would have been formed. Even spirochetes have their role in this endosymbiotic theory: they could have contributed their motility system to the origin of complete unicellular organisms moving with the aid of flagellae and cilia.[S23,G27] Furthermore, the same motility system in mitochondria-containing prokaryotes may have facilitated the formation of a real nucleus in which the genetic material is arranged in chromosomes and performs complex movements during cell division (Fig. 3.1).

According to endosymbiotic theory, eukaryotic cells themselves are also composite living organisms—several organisms co-operate to form a single system living on a higher level, such that the lower-level systems retain their own life within the higher unit. It is worth mentioning that it is not difficult to find recent examples of a similar symbiosis. Unicellular organisms often absorb algae which are not digested but remain living within the cell, functioning and multiplying. If the 'host' cell divides, the algae statistically divided between the two descendant cells. Similarly, organisms which were formerly thought to be self-contained photosynthesizing systems were found to stop photosynthesizing if they were cultured in the dark; after a number of divisions the descendants no longer contain any chloroplasts and thus are unable to photosynthesize. However, they are still capable of living and multiplying in the presence of nutrients.

Clearly, we cannot look for the elementary units of biology—analogous to points in geometry, mass-points in mechanics, elementary

[S22]Mitochondria and plastids are certainly not alive. Most of their essential genes reside in the nucleus of the host cell and they must import proteins from the cytoplasm to ensure their own functioning and division (Alberts *et al.* 1994). The *ancestors* of these cell organelles *were living bacteria* more than a billion years ago.

[S23]There is no convincing evidence in the recent literature to support the symbiotic origin of cilia.

[G27]The spirochete endosymbiosis theory is discussed by Margulis (1981).

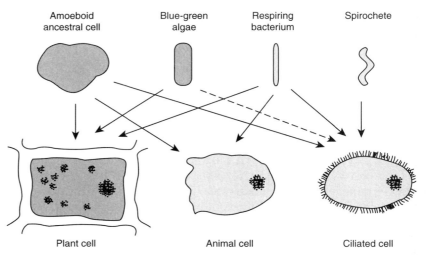

Fig. 3.1 According to our present knowledge more complex cells (plant, animal, and ciliated cells) originated from the symbiosis of simpler cells without nuclei—the prokaryotic cells.

charge in the theory of electricity, and molecules in chemistry—at the organizational level of multicellulars organisms or even of eukaryotic cells. The basic units of theoretical biology must be sought at the organization level of prokaryotes.

However, prokaryotes themselves are quite complex systems, even though they cannot be separated into living parts. The prokaryotes known to us are not identical with the simplest primeval living systems; they are the product of an evolutionary process measured in thousands of millions of years. It is well known that the internal processes of prokaryotes are regulated and controlled by extremely complex regulating systems on several levels. It is sufficient to consider the results of molecular biology: The synthesis of DNA, the transcription of genetic information to RNA, the mechanism of protein synthesis, and the determination of the amino acid sequence have all been studied on bacteria, i.e. on prokaryotes, and even today most of our knowledge is restricted to prokaryotes.

The enzymatic mechanisms which control and regulate the functioning of even these most simple living organisms are extremely complex. Molecular biology looks for the presence or absence of enzymes behind the individual properties, and if an enzyme which controls a particular biochemical process is not present in the cell, this process cannot occur. At the present level of the living world, this view is obviously correct.

Chemistry appears to contradict this view. As we know, enzymes are catalysts. According to chemistry, catalysts can only accelerate chemical processes which occur naturally, although far more slowly in the absence of a catalyst. However, the two views, can be considered to be consistent for the following reason. Enzymes can increase the rate of a chemical reactions by a factor of 10 to 100 million, and, from a biological viewpoint a process within a cell which runs 10 million times slower without an enzyme can clearly be considered as not to be occurring at all.

In the living organism, even the simplest prokaryotic cell, many chemical reactions run simultaneously. A given chemical substance is not restricted to a single chemical reaction; the various reactions are connected and form a complicated unified reaction network. For a biologist, this network is determined by the enzymes present in it. For a chemist, the basic properties of the network are determined by the transformations of the participating chemical compounds; the enzymes present in the cell merely determine the end result.

Thus the role of the intricate enzyme systems in the cell is only to control and regulate. Regulation is performed by increasing or decreasing the rate of partial functions in a complex system which has fundamentally characteristic properties. Hence the basic characteristics of living systems are not determined by the enzymes; these merely speed up or slow down events, which means that they may obscure the properties of the regulated system.

In looking for the fundamental elementary units of biology, we must not be led astray by the efficiency of the regulation carried out in them. The truly fundamental basic units of biology should not be sought in the enzymatic regulation, but in the system regulated *by* the enzymes.

3.3   Sets and systems

Language with its subtle devices is often capable of exquisite differentiation of the phenomena of the surrounding world. Science with its exact methods reaches into categories which attempt to define the endless diversity of the world. However, in order to name these categories, science has to use words from the common language, defining their meaning accurately although sometimes rather arbitrarily.

Scientific and everyday interpretations do not usually coincide. Scientific interpretation is artificial, while the common linguistic interpretation is the result of a long refinement during many centuries of use and thus often reflects the true situation more exactly and more subtly. This also holds for the expressions 'set' and 'system'.

In normal usage the word 'set' refers to a disordered manifold, to some group of things. If we speak about a set of bricks, we do not assume that it will be ordered in any way. On the other hand, the word 'system' somehow suggests the idea of order and regularity. Things are quite different in scientific practice. In set theory, a rather new branch of mathematics, any manifold of things (elements) is considered a 'set', regardless of the amount of order within the manifold. Similarly, in thermodynamics, a highly important discipline of physics, every part of the world limited by real or imaginary walls is considered a 'system', again regardless of whether order or disorder rules within the walls.

Cybernetics and system theory are two of several disciplines developed in the last decades. Cybernetics deals with the functioning of dynamic systems, and system theory with all types of system. But both of them deal only with systems within which there is some kind of order, i.e. the individual parts of the system—its elements—are in a well-defined organizational relation. Thus the concept of system in cybernetics and system theory is narrower than that in thermodynamics.

A number of definitions of 'systems' can be found in textbooks of system theory and cybernetics, but all of them are coined from some special viewpoint and, not infrequently, contradict each other. In biology, order is of fundamental importance. It is not by chance that one of the founders of system theory, Ludwig von Bertalanffy, was a biologist. During his investigation of the organizational laws of living systems he discovered quite general relations of organization, which are valid not only for living organisms but also for everything which has a genuine internal order and organization.

In seeking the basic laws of living systems, we first have to agree the interpretation of the basic terms of these systems. These terms are *not* used in the same sense in the various scientific disciplines. For biological purposes, all these concepts have been formulated anew, regardless of their interpretations in other disciplines. However, the following formulations regarding systems also appear to be valid in other non-living domains of nature.

First, we have to define the word 'set'. It implies a collection of a number of things (elements) regardless of whether they are in an ordered or a disordered state. Thus sets can conveniently be divided into two large groups: disordered and ordered sets or systems. Disordered sets are not dealt with in the following. Some systems, (e.g. a pile of bricks or a military marching column) are governed by geometric order. In other systems order is not so obvious, but it is still clear that the parts are related to each other in same way and are not a completely random collection. Examples of this kind of order are a machine or a radio. Yet other kinds of systems have connections which can only be recognized in their functioning over a period of time. A photograph of the solar system reveals a disordered set of celestial bodies; a photograph of a beehive or an ant-hill displays a confused mass of bees or ants. However, if the functioning of these sets is observed over a period of time, their organization, i.e. their property of being a system, becomes obvious.

The systems in the first group can be called 'geometrical', since they differ from disordered sets because of the geometrical symmetries appearing in them. However, these do not interest us any further. The two other groups form dynamic functioning systems which display qualitatively new operational properties whilst functioning. However, these properties depend on inner organization and will disappear when it is absent. Therefore these are dynamic systems. In one of these two groups of dynamical systems, ordered functioning is guaranteed by a geometrical structure (but not symmetry!) of fixed materials; in the other group, the elements of the system are not connected by a geometrical structure of fixed materials but interact through space, and thus these systems are modifiable and have a soft geometrical structure. Thus the first type can be called hard systems, and the second type soft systems.

The ant-hill (or more exactly not the 'hill' as a structure but the community of ants) is a dynamic soft system, since regulation and control of its operation occurs 'through space' by interactions between its elements—the ants. These interactions, which are produced by pheromones (chemoattractive substances) and other individual contacts (e.g. signs made with the antennae), rather than a solid geometrical structure of parts and connections, control the functioning of the system.

The elements of the ant-hill, the individual ants themselves, are systems on the next-lowest level of organizational hierarchy, and are dynamic. But are they also soft systems? They have a chitin cover of a definite and characteristic geometrical form, and their body parts are integrated by 'constituents' and 'connections' which regulate and control their movements and actions. But this is only a part of the picture. Their actions and their development are also regulated by specific hormones, which act at a distance, 'through space', and their life functions are also controlled by the soft processes of their metabolism. Moreover, during pupa formation the geometrical structure of the larva dissolves and a new structure is formed so that the life of the individual ant remains intact. The life of the ant-individual certainly depends on its inner organization, but not so much on the geometrical as on the 'soft' structure of this organization. Similarly, the caterpillar and the beautiful butterfly which develops from it have the hidden 'soft' organization and not the geometrical structure in common (Fig. 3.2). Organisms are fundamentally soft systems, from molluscs like medusae and octopi to the hardest trees such as oaks and nut trees.

Ants are metazoans. The basic elements of their systems are cells, while on the next-lowest level of organizational hierarchy cells themselves are systems—dynamic soft systems. Cells are living systems. Each cell has a distinct self-reliant life; however, its function is subordinated to the

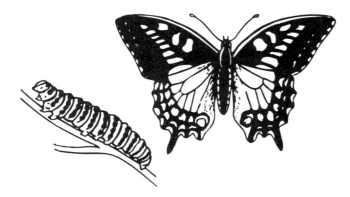

Fig. 3.2 The hidden 'soft' organization and not the geometric structure is the common property of the caterpillar and the beautiful butterfly which develops from it.

functioning of the ant as a system on the next-highest level, in the same way as the activity of the ant depends on the 'life' of the ant-hill.

The ant is living, and the cells are also living. The ant-hill can be demolished but the individual ants remain alive; individual ants can be destroyed but the ant-hill remains alive. Similarly, if the ant is killed, its cells may remain alive and, conversely, individual cells can be destroyed but the ant remains alive. Thus the life of the ant is not identical with the life of its cells or with the 'life' of the ant-hill. Each of these systems lives on a different organizational level. A million ants are not an ant-hill, nor do a thousand million ant cells form an ant. The ant as an ant is only determined by its inner organization—the soft organization connecting the functioning of the cells. Similarly, the ant-hill as an ant-hill, i.e. a 'living' dynamic system, is determined by the soft organization connecting the activities, i.e. the functioning, of the individual ants. Now what makes the cell living? The soft organization of its inner events and occurrences. Thus, if we are looking for the fundamental laws, for the principle of life, we have to establish the connections of this soft organization.[G28]

It was mentioned above that an ordered pile of bricks is a system. If this is halved so that the order of the bricks is not disturbed, both of the resulting two halves remain systems. Their quantitative properties change, but the qualitative properties of the original and the resulting systems are the same.[G29] A crystal of rock-salt shattered by a hammer breaks into many small pieces, but the small crystals have the same qualitative properties as the original rock-salt crystal; symmetries, angles between faces and edges, optical properties, etc. are all identical. Similarly, the railway network of a continent is a system. If a war breaks out on the continent, this system will break into two or more parts—two or more independent smaller systems. But each subsystem continues to function, and each displays the qualitative properties of a railway.

A radio is also a system. However, if a radio is cut into two or more parts, its qualitative properties inevitably change; we cannot divide a radio so that both parts remain radios. Similarly, an automobile is a system, but an automobile cannot be divided so that both parts remain automobiles. The ant is a system, but the ant cannot be divided so that both parts remain living ants.

The first type system is called a divisible system, and the second type is called a unitary system. Living systems are obviously unitary systems. Divisible systems are generally composed of several, sometimes very many, unitary systems which are connected in a complicated manner such that they can be merged into or covered by each other.

The basic laws of life should be sought in the organizational mode of the simplest biological unitary systems, i.e. at the level of prokaryotes which are the simplest cells without nuclei, mitochondria, chloroplasts, or flagella. However, even the concept of the unitary system is not sufficiently clear to sustain the fundamental principle of life. This is illustrated by the following example.

[G28]This kind of argument raises problems similar to those discussed by Hobbes in his story of the ship of Theseus. Theseus had a ship which he repaired from time to time by replacing worn-out parts that he discarded on the beach. Unbeknown to him, someone else collected the discarded parts. The scavenger eventually built a complete boat out of the parts discarded by Theseus. Did the scavenger thus acquire Theseus's ship, or did Theseus still have his ship? If the latter, then why is the scavenger's ship, which is entirely constructed of parts that formerly made up Theseus's ship, *also* Theseus's ship? The metaphysical property of being a thing, identity conditions for things, and related philosophical problems are discussed by Jubien (1997).

[G29]Wimsatt (1985) discusses various kinds of property invariance under operations like halving in terms of an analysis of various forms of aggregativity (like piles of bricks). Failures of aggregativity describe many aspects of organization and emergence. Wimsatt (1974) characterizes biological organization and several senses of complex organization in terms of criteria for the 'near decomposability' of systems into parts.

The case enclosing a radio can be removed without any effect upon its sound and, similarly, many other parts can be removed without changing its fundamental qualitative properties. The lamps, body, and many other parts of an automobile can be removed but it still remains an automobile. Thus these systems have some parts which can be removed without influencing their fundamental qualitative properties. However, it is not possible to remove a single atom from an acetic acid molecule and still retain the properties of acetic acid.[G30] If we replace one oxygen atom with two hydrogen atoms, we obtain an ethanol molecule with entirely new qualitative properties. If we then remove a hydrogen atom we obtain acetaldehyde, again with completely different qualitative properties.

This is why, in the previous section, molecules were defined as the basic elementary units of chemistry. Similarly, the basic elementary units of crystallography were the elementary cells consisting of several atoms, ions or molecules, arranged according to the three types of crystal. No atom, ion, or molecule, i.e. no building element, can be removed from these elementary cells without destroying the characteristic crystallographic properties. There is nothing superfluous in these systems; they are constructed from the minimal number of elements, i.e. subsystems, necessary for the existence of the given qualitative properties. Therefore we shall call these modules minimal systems. New qualitative properties are the result of the organization of minimal systems!

In section 3.1 we saw that the exact sciences are based on the basic elementary units of their own domain or, more exactly, on the abstract theoretical models of these units. Now we can refine this formulation by stating that exact sciences are based upon abstract models of their own minimal systems. And now it is clear why—the qualitative properties characteristic of the discipline in question appear first in these minimal systems.[S24] Thus the atom (the hydrogen atom) displays the basic qualitative properties of nuclear physics, the molecule those of chemistry, and the elementary cell those of crystallography. The new qualitative properties can be understood most readily in these minimal systems, where they can be dealt with most exactly, and the models of these minimal systems can be formulated with absolute mathematical precision. The minimal system of the generator is a loop of wire rotating in a magnetic field, and that of a radio is a 'detector' radio consisting of a valve, a coil, a condenser, and a receiver.

The principle of minimal systems is in a sense independent of their actual existence. The molecule as a minimal system of chemistry actually exists and is stable, the wire loop rotating in a magnetic field serves only as a teaching aid, and the elementary cell of crystallography does not exist in alone, as it is unstable and is only stabilized by the presence of millions of similar cells within the crystal. The appearance of new qualitative properties is still bound to these minimal systems independently of their actual existence, and the scientific treatment of properties depends on the models of the minimal systems and on the exact description of their quantitative relations.

[G30]This claim depends as much on the properties of interest to chemists as it does on the properties of molecules. If the outside casing of a radio is removed, the property of being a radio *of a certain kind* is not preserved, even though the radio may still function. Acetic acid with certain atoms removed still retains some chemical properties in common with acetic acid, for example being a molecule containing at least one carbon atom if only a hydrogen atom is removed. Substituting one side chain of an amino acid (e.g. CH_3) by another (e.g. $CH(CH_3)_2$) does not remove the property of being an amino acid. In this case, it only changes alanine to valine. Both are hydrophobic amino acids, and so hydrophobicity is preserved by this substitution.

[S24]The situation is more complicated in biology than in other sciences for the following reasons: (i) the complexity of organisms can develop through evolution; (ii) because of evolution, the simplest living systems today are unlikely to be minimal systems in the logical sense.

S25There are organisms that are *not* unitary, but modular. They are usually divisible, and can thus regenerate the missing modules easily. Whether a modular organism can be regarded as a single living unit, or merely a network of living units (which seems to be Gánti's preferred option), is open to further discussion.

As has already been mentioned, organisms are unitary systems.[S25] However, there is not a single living system in the present living world which is a minimal system! Just consider, how many things can be removed from a man without robbing him of his life or of his manhood! Mutagenic effects may modify or destroy many properties of cells without killing them. Present-day living systems have many accessory parts, properties, and capabilities which are not essential to life itself, but are only necessary to a life which is differentiated, qualified, and refined within the community of the living world, in given environmental conditions, by the present state of evolution.

However, in our search for the basic principles of life we have to consider the simplest living system—a minimal system which already displays the properties characteristic of life as 'life' and from which nothing can be removed without the loss of these properties. This minimal system will be the ultimate elementary unit of life, and may be the fundamental unit of an exact theoretical biology. In the following sections we shall consider this ultimate elementary unit of biology—this minimal system—and the laws governing its functioning, and we shall try to deduce from these minimal systems the phenomena of the living world. We shall call these final elementary units chemotons, as described in the first part of this book. First, however, we must review the general properties and governing laws of dynamic systems.

3.4   Function and stability

Language divides changes into two groups: occurrences and functions. An occurrence is a change which occurs once; for example, a material system passes from one state into another. Function means a permanent, yet preserving change; the material system can undergo permanent changes but actually remains unchanged. Is this a contradiction?

An explosion in a quarry is a single process which cannot be repeated—the explosion occurred. If the gas injected into the cylinder of an internal combustion engine is ignited, an explosion also occurs, but this process can be repeated many times. The explosion can be performed in the same manner and with the same consequences many millions of times. The internal combustion engine functions.

In both cases chemical energy is released, and this energy, or at least a substantial part of it, is used to perform work. In the former case work is performed once, but in the latter it is performed continuously. Continuous work performance can only be achieved by means of suitable work-performing systems characterized by changes occurring through a series of constrained motions, such that the inner organizational characteristics of the systems remain unchanged.

Therefore there are two conditions for continuous work performance: one is the released energy which can be changed into work, and the other is the system performing the work. Let us consider first the energetic conditions of work performance.

Our view—and usually the view of physicists—is governed by the mechanical concept of work, according to which work is equal to the product of the force and the displacement occurring in the direction of its action. Therefore if no force is applied, no work is done. Nor is any work done if there is no displacement in the direction of the acting force. However, we now encounter a contradiction which is shown in the following example.

Suppose that during the building of an iron bridge a structural element must be held for 15 minutes at a height of 2 m in order to perform the necessary mountings, fitting and tightening of screws, etc. In principle, the task can be performed in several ways, which are considered below. It must be emphasized that raising the structure to the required height necessitates the investment of work, but it can be maintained at this height with or without performing work (Fig. 3.3).

The first solution is to build temporary scaffolding which supports the heavy structural iron element during the assembling. In this case maintaining the structure at a height of 2 m does not require work, since the gravitational force acting upon the structure is compensated by a force of the same magnitude but of opposite direction produced by the mechanical strength of the timbering. In this case we are dealing with a mechanical equilibrium.

The second solution does not need any timbering as the iron structure is held in place for 15 minutes by a sufficient number of workers. According to the physicist, no work is performed as there is no displacement. However, the workers will think that they are performing hard physical work, and indeed biological tests show that the same phenomena are registered in their bodies as when they are performing physical work in the 'mechanical sense'.

In the third solution the structure is attached to a balloon filled with a gas lighter than air. No work necessary to hold the iron structure, apart from that expended in manoeuvering it into position, as the gravitational force acting upon it is compensated by the buoyancy of the balloon.

In the fourth solution the balloon is replaced by a helicopter. Clearly, the helicopter has to perform additional work while it holds the iron structure in the air, which is revealed as an increase in fuel consumption. The surplus mechanical work performed can be calculated from the increase in the amount of moving air etc. Thus in this case mechanical work is obviously required to hold the iron structure at the given height.

In summary, in the first and third case no work is required to hold the iron structure at a height of 2 m, in the second case some kind of 'biological' work is necessary although no physical work is required, and in the fourth case continuous mechanical work is required.

In order to solve this paradox, we must recall some basic theorems of physics. The first of these is that every material set or system spontaneously strives to reach a state of equilibrium. However, many kinds of equilibrium states are possible: an object in mechanical equilibrium does not involve thermal equilibrium, a chemical equilibrium does not involve a magnetic or electric equilibrium, etc. In the above

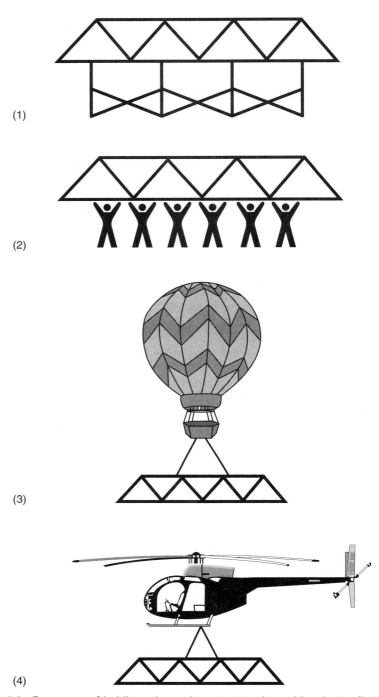

Fig. 3.3 Four ways of holding a heavy iron structure in position. In the first and third cases no work is performed and, in the fourth case permanent mechanical work is required. But what about the second case? From the viewpoint of the physicist, no work is performed. However, the workers holding the iron structure are unlikely to believe this.

example we were dealing with a mechanical equilibrium, which exists when the vectorial sum of the mechanical forces acting upon the object is equal to zero. This condition is satisfied in the first and third cases. In the first case, the gravitational force acting upon the object is compensated by the forces developed by the deformations caused in the timbering. In the third case, it is compensated by the buoyancy caused by the difference in specific weights.

However, a static mechanical equilibrium cannot occur in the second and fourth cases because there is no compensating static force. Thus in these two cases the system cannot be brought to a state of equilibrium mechanically. Since, as mentioned, every material set or system strives to reach a state of equilibrium, a material set or system in a state far from equilibrium can only be maintained by some permanent energy investment, i.e. by the continuous performance of work. In the case of the helicopter this appeared as mechanical work. However, when the iron structure was held up by workers, the work was chemical, not mechanical.

The mechanical definition of work is not sufficient to completely understand chemical work. According to the thermodynamic definition of work

$$-\Delta U = -(Q + L)$$

where ΔU is the change in the internal energy of the (thermodynamically defined) system, Q is the heat produced, and L is the work performed. (The negative sign refers to the fact that the system loses internal energy by emitting heat and doing work.) Thus in this definition work is related to the change in energy and the heat produced rather than to displacement and force.

Work is always performed through some system which, theoretically, directs the released energy on a constrained path and does not allow complete conversion to heat. For instance, the energy released in an oil lamp is almost entirely in the form of heat, but in an internal combustion engine most of the energy released is transformed into mechanical energy.

In machines, chemical energy is transformed into mechanical energy, and the useful work is performed by some mechanical means. In accumulators and batteries chemical energy is transformed directly into electric energy. In the chemical industry and in chemical laboratories chemical energy is often transformed into chemical energy, i.e. the decomposition of one compound is used directly to synthesize another compound. This also involves the performance of work, but here it is chemical work.

The fundamental form of work in living organisms is chemical. The cells 'break down' the nutrients, i.e. they transform them into compounds with a lower internal energy. The energy released, apart from the inevitable heat losses, is used to synthesize the constituents of the cells, thus ensuring their continuous growth and reproduction. Some cells can also use chemical energy to perform mechanical or electrical work, or to emit light. Thus in our example, work was done when the workers held

Fig. 3.4   The windmill changes permanently as it performs work but its final state is unchanged from its original state.

the iron structure motionless, since in order to balance the gravitational force they had to strain their muscles at the price of a performing continuous chemical work.

We have mentioned that continuous work can only be performed by using a system which changes during work in such a way that it finally remains unchanged. In an internal combustion engine the explosion moves the piston from its original location, but the engine is so constructed that the displacement occurs on a constrained path and after performing work the piston returns to its starting position. The windmill (Fig. 3.4), the water turbine, the electromotor, the lathe, and the planer all move on constrained paths, and their continuous function can be broken down into a chain of sequential work-performing cycles. This cycling behaviour, or periodicity, can be seen directly in mechanical work-performing constructions. The ability of non-mechanical systems to perform continuous work also depends on cyclic processes or, as they are often called for simplicity, cycles. A well-known non-mechanical work-performing system is the refrigerator, where a cooling gas in a work-performing system passes through a sequence of cycles of different states of pressure and temperature.

Living systems are dynamic functioning systems. Life itself is the continuous organized functioning of the system, which can only be maintained at the price of continuous performance of work.[G31] Thus living organisms are never in equilibrium; the living state is characterized by the ability of living organisms to maintain themselves permanently far from equilibrium, at the price of a continuous expenditure of energy. Here again we have a paradox, because of the generally accepted usage of words.

[G31]In other words, living systems are dissipative structures (Maynard Smith 1986).

In everyday usage, equilibrium and stability are identical or nearly identical concepts, in the sense that the properties of a set or system do not change with time in a state of either equilibrium or stability (or at most they fluctuate about the equilibrium/stability state). However, in thermodynamics the concept of equilibrium is restricted to the stable state of closed systems. From this it follows that living systems cannot be in a state of equilibrium as they are open systems through which a stream of material flows during their metabolism, as this is one of the fundamental phenomena of life.

However, open systems can also have (stable) states in which the properties of the set or system do not change with time, namely if the quantities of the material coming in and going out in unit time are the same. This state is said to be a stationary or a steady state. The theoretical bases of the stability of stationary states were being established in irreversible thermodynamics, at the time when Bertalanffy's organism-concept was becoming widely used. Bertalanffy also considered the stationary state as the basis of the stability of living beings, and since then this opinion has become widely accepted. Moreover, the stationary state is sometimes considered to be identical with homeostasis.

Incorrectly, of course. The stationary state is, by definition, a state of open systems with an equal rate of inward and outward movement of matter. However, living systems are fundamentally growing (accumulating) systems, in which more matter enters than leaves. A growing system cannot be in a stationary state, and hence attempts to reduce the stability of living systems to the irreversible thermodynamics of open systems in the steady state are *ab ovo* doomed to failure.

Obviously, stability is not merely a mechanical or a thermodynamical problem. It is a basic problem of almost every system, ranging from natural material systems through man-made contrivances to economical and political systems. Understandably, the question of stability appears in the most diverse disciplines; thus stability tests are applied to economic systems, and stability criteria are used in control and automata theory. 'Cybernetic' stability criteria were developed by Ashby, who even constructed a 'cybernetic homeostat' which, contrary to the common belief, does not produce homeostasis, since homeostasis is the 'equilibrium' state of accumulating systems whereas the Ashby homeostat is not an accumulating system.[G32]

[G32]See Ashby (1952).

Stability criteria have also been formulated using abstract mathematical methods. This was done at the end of the nineteenth century by the Russian mathematician Lyapunov. To this day, most mathematical stability investigations are based on Lyapunov's work (e.g. the Lyapunov functions for the system in question). However, Lyapunov stated in his fundamental principles that a system with permanently increasing or decreasing energy cannot be in a stable state, and thus accumulation systems are *ab ovo* excluded from the domain of Lyapunov stability. Accumulation, growth, and multiplication are fundamental to the living world, whether we consider prokaryotes, eukaryotes, multicellular organisms, or populations. But all these systems maintain their 'internal constancy', or homeostasis, during growth and compensate for adverse changes in the external environment by dynamic responses. In living systems (cells and organisms) this compensation occurs via the same fundamental biochemical mechanisms that use the chemical energy of the nutrients to perform effectively directed work.

It was stated above that changes in work-performing systems must occur along a series of constrained paths. The compensation of external effects can occur only by some kind of regulation, which also demands that changes occur along a series of constrained paths. Finally, growth and multiplication require that the processes should all be controlled, which also presupposes a series of changes occurring along constrained paths.

The presence of constrained paths in mechanical devices is accepted as something quite natural. We encounter such constrained motions every day, from the swinging of a pendulum to the cycles of a piston. In electric and electronic devices changes in the current are invisible, but we know that the wires are the constrained paths of this change and that if they are short-circuited the device fails to work. However, in living systems the transformation of energy, performance of work, regulation, and control all occur by chemical means in solution. Where are the constrained paths, and what are they like?

Clearly, in looking for the secret of life—the fundamental principle of life—we have to examine the nature and organization of 'constrained paths' involving of chemical changes capable of work performance, control, and regulation in solutions.

In 1966 the author stated that certain chemical cycles and their more complex forms—reaction networks composed of interlocking cycles—are the basis of life processes in all living systems and given in the stability and homeostasis of living organisms. The 'chemical motor' transforming chemical energy into directed work, i.e. the 'functioning' of the living system, consists of chemical reaction cycles. The very cyclicity of these cycles ensures the variability and stability of living systems. Despite continuous change and function, from an organizational point of view the system is always the same. This stability ensured by cycles is totally consistent with the accumulation processes (growth and multiplication) as will be shown in the discussion of the organization of chemotons.

3.5   The criteria of life

Living and non-living systems are qualitatively different, i.e. living systems have qualitative properties or groups of qualitative properties which occur exclusively in the living world and cannot be found in the non-living world. In what follows, these common characteristics found in living organisms will be called life criteria, the laws uniting these characteristics will be considered the principle of life, and life itself as a common general abstraction of every kind of living being will be accepted as a philosophical and not a biological category. Life criteria will be dealt with in the present chapter, and the principle of life will be discussed in connection with the organization of the chemoton. Life as a philosophical category will not be dealt with in the present book.

The selection and axiomatically exact formulation of life criteria is of fundamental importance in theoretical biology. As we have seen, classical biology has not been able to solve this problem during its long history of 2000 years.[G33] We shall present here a quite new system of life criteria, which entirely differ from the classical 'life phenomena'. This new system of life criteria was first proposed in the original edition of this book in 1971. However, in the diverse world of biology it is not easy to select fundamentally common characteristics, since it is impossible to find a research worker who could claim to know every part of the living world with the required precision. Thus the original system of criteria has been somewhat modified since 1971, and here we present a refined and updated form, including both content and formulation. Nevertheless, even this formulation cannot be considered the final version, either in content or in axiomatic exactness, although it is unlikely that major changes will be necessary in the future.

Life is a function of material systems which are organized in a particular way. Thus life is not the property of a special type of matter in the sense of chemistry, i.e. of a chemical compound (e.g. protein or nucleic acid), but is the property of specially organized systems. Thus it is incorrect to speak about living matter; it is more appropriate to speak about living material systems.

A system is living if, and only if, certain characteristically composed processes (life processes) occur in it. The totality of these processes, i.e. the functioning of the system, results in characteristic phenomena which are well suited to differentiate the living from the non-living.

A system suitable for the occurrence of the processes in question may be either functioning or in a state which is not functioning but is capable of doing so. When this system is in its functioning state it is said to be living; when it is in its non-functioning but functionable state it is capable of living but it is not dead. This latter state corresponds to latent life, clinical death, resting seeds, dried-out micro-organisms, and frozen organisms. This is a state which is non-living but not dead.

Death is an irreversible change which makes the system irreversibly incapable of functioning. Consequently, if life corresponds to the functioning state of these special systems, death corresponds to the state which is incapable of functioning. However, there is also an intermediate state which is capable of functioning but is not actually functioning, i.e. a state which has the capability of life but in which the system is not alive because the special processes are not occurring, but it is not dead either because the processes can be started if suitable circumstances arise.

The totality of life processes, i.e. the functioning of living systems, produces special phenomena which are appropriate for the general characterization of the living state. Some of these phenomena are present in every living being, without exception, at every minute of its life. Without the constant presence of all these phenomena, the system cannot be living. The presence of all these phenomena is an unavoidable criterion

[G33]Formulation of life criteria goes back at least to Aristotle. A contemporary philosophical assessment of life criteria and the strategy of understanding life in terms of them is given in a very useful book by Feldman (1992). He argues that a wide variety of life criteria fail to serve as analyses of life, i.e. to serve as proper definitions of life. However, Feldman's assessment depends on his particular reading of criteria such as Aristotle's 'nutrition', which he compares to contemporary requirements of 'metabolism'. Gánti's view of metabolism is differently focused—on the biochemical processes rather than on the mere intake of food—and so his account is not vulnerable to Feldman's argument that no life-functional analysis of life succeeds.

of the living state; therefore they will be called real (actual, absolute) life criteria. However, there is another group of life phenomena whose presence is not a necessary criterion for individual organisms to be in the living state, but which are indispensable to the survival of the living world. These will also be considered life criteria, and they will be called potential life criteria.[G34]

3.5.1   Real (absolute) life criteria

1. A living system must inherently be an individual unit

A system is regarded a unit (a 'whole') if its properties cannot be additively composed from the properties of its parts, and if this unit cannot be subdivided into parts carrying the properties of the whole system.

A system forming a unit (unit system) is not a simple union of its elements, but a new entity carrying new qualitative properties compared with the properties of its parts. These new properties are determined by interactions occurring according to the organization of the elements of the system. Only the system as a whole displays the totality of these properties.[G35]

Living systems form inherently, i.e. by their very existence as a unit. Thus living systems are inherently unit systems—life always is the property of a unit system. However, the inherent unity of living systems does not exclude the possibility of accessory components, i.e. a unit system does not have to be a minimal system at the same time. Since living systems are genetically determined structurally as well as functionally, the biological unity of the system is also genetically determined. The genetic program also carries information with respect to the living unit.

2. A living system has to perform metabolism

By metabolism we understand the active or passive entrance of material and energy into the system which transforms them by chemical processes into its own internal constituents. Waste products are also produced, so that the chemical reactions result in a regulated and controlled increase of the inner constituents as well as in the energy supply of the system. The waste products eventually leave the system, either actively or passively. Expressions such as 'external' and 'internal' do not refer here to some spatial separation, but allude to the question of whether or not the material is an organic part of the internal organization of the living system as a unit system. Thus stored nutrients such as glycogen or starch are considered as external materials even if they are spatially located within the living system.

3. A living system must be inherently stable

Inherent stability is not identical with either the equilibrium or the stationary state. It is a special organizational state of the system's internal processes, which makes the continuous functioning of the system possible and remains constant despite changes in the external environment.

[G34]Another tradition which developed independently of Gánti's work, which distinguishes primary or 'real' criteria for living organizations in contrast with the criteria for a living world, is the theory of autopoiesis (Maturana and Varela 1980).

[G35]This argument should be taken to apply to just those entities with qualitatively new properties that *do* result from interaction among the parts of a system, i.e. where this is an additional criterion required for being a unit. The text suggests that this qualitative newness follows from the non-additivity and non-subdividability criteria given. It is trivially true that every simple union of elements (physical parts)—what philosophers call their 'mereological sum'—is a new entity carrying new qualitative properties. For example, the entity consisting of Gánti's pencil, Szathmáry's computer, and Griesemer's telephone has the qualitatively 'new' property (i.e. different from any property of the parts) of being the sum of just those three things. It also has the property, which none of the parts does, of being distributed over two continents.

It means that the system as a whole, although continuously reacting via dynamic changes occurring within the living system, always remains the same. Further, it means that despite the permanent chemical transformations occurring in the living system, the system itself does not decompose; rather, it grows if necessary.

Inherent stability is something more than homeostasis, since homeostasis follows from it. Inherent stability is an organizational property—a natural consequence (as we shall see later) of the network of chemical and physical processes taking place in the living system. The dormant seed, the frozen tissue culture, the lyophilized micro-organism and the dried-out protozoon are not in homeostasis or in a stationary state, although all of them satisfy the criterion of inherent stability, since under appropriate circumstances they can live again. Thus a system in a state of inherent stability only displays homeostatic properties whilst functioning; hence this criterion includes that of the homeostasis. Thus a living system with inherent stability is capable of living, but does not display homeostasis in a non-living state. Under appropriate circumstances it can be converted from a non-living to a living state, and then it acquires the property of homeostasis.

The concept of homeostasis was proposed the great American physiologist, Cannon. Living organisms have a special 'inner environment', differing in its state and composition from the external environment. Living systems, whether they are simple cells or complex multicellular organisms, strive to hold this inner environment constant despite external changes. Cannon called the constancy of the inner environment homeostasis. As has already been seen, this constancy of the inner environment is also a property of growing and multiplying systems. Thus homeostasis cannot be identified with equilibrium, the stationary state, Ashby's 'cybernetic homeostat', or Lyapunov stability.

The constancy of a living system's inner environment, i.e. homeostasis, can only be maintained by detecting the changes occurring in the external environment and reacting to them with active compensating answers.[G36] Thus the mechanism of homeostasis is excitability. This can also be carried out in soft molecular systems, as will be seen in connection with the chemoton theory.

Thus inherent stability as a life criterion includes the criteria of homeostasis and excitability in the most general sense, so it is unnecessary to list them separately.

4. A living system must have a subsystem carrying information which is useful for the whole system

Everything carries within it the information necessary for its origin, development, and function.[G37] However, there are some systems which carry information concerning things and events which are independent of them. Examples of such systems are books, magnetic tapes, punched cards, and discs.

In nature the capacity to carry such 'surplus information' can only be found in certain subsystems of living systems such as the genetic

[G36]Waddington called this kind of dynamic stability of process, in which a trajectory of change rather than a particular state is maintained, 'homeorhesis' ('The basic ideas of biology', 1968; reprinted in Waddington (1975)).

[G37]Everything carries some, if not all, information necessary for its origin, building up, and function. It is not clear that things generally carry all the information necessary for any of these. For example, a sculpture does not carry the information about how it was or is to be made. If it did, we would not need a Rodin or a Michelangelo to make a good one. Nor does a bottle opener carry all the information necessary to its ability to open bottles, since some of the information is in the bottle in the shape and position of the cap, the sort of leverage point on the bottle needed to operate the opener, and so forth.

[G38]Just as Gánti has argued that life is a certain kind of organization rather than a special kind of matter, I would argue that genetic properties are properties of a certain kind of organization rather than of a special kind of matter. Thus it is misleading to speak of a 'genetic substance'.

substance, brain, immune system, etc.[G38] These carry not only information about themselves, but also information about the whole system or even the world outside the system. The presence of information-carrying subsystems is characteristic of every living system without exception, and is an indispensable criterion for the development of the living world.

Information coded in a system becomes real information only if there exists another system capable of reading and using this coded information. Living systems can read and use the information stored in their information-carrying subsystems; moreover, they can copy, i.e. replicate, information during their multiplication. Living beings are characterized not only by their ability to store information, but also by their ability to carry out informational operations.

5. Processes in living systems must be regulated and controlled

An existential condition of every dynamic, i.e. continuously functioning, system is the regulation of its processes. Living systems as soft dynamic systems also have this property. Regulation in living systems occurs primarily through chemical mechanisms. In fact, it could be claimed that regulation should not be a separate criterion, since neither metabolism nor homeostasis can be realized without regulation of the system's processes, and therefore it is already implied by these criteria.

Nevertheless, regulation itself can only ensure the maintenance and functioning of the system. Unidirectional irreversible processes, such as growth, multiplication, differentiation, development, and evolution, also occur in the living world and these cannot be accomplished by regulation alone. They also require steering, i.e. control according to some program. Control in living systems is also performed by molecular mechanisms.

3.5.2 Potential life criteria

1. A living system must be capable of growth and multiplication

Growth and multiplication are among the classical life criteria; their presence is general and indispensable in the living world. However, they are not criteria of the living state itself: some domesticated plants and animals are *ab ovo* incapable of reproduction; castrated animals are unable to reproduce; old animals lose their ability to grow and reproduce without losing their lives. The presence of growth and reproduction is not a criterion for individual life, but it is a condition for the existence of the living world and so it must be included among the potential life criteria.[G39]

[G39]This view that reproduction is a potential rather than a real life criterion is shared by the autopoietic tradition.

It is necessary to explain why growth and reproduction are combined in a single criterion. Life criteria must be valid for every individual of the living world on any of the levels of the organizational hierarchy. However, growth and multiplication separated only at a given level of evolution; in prokaryotes and in many eukaryotes growth is merely a part of reproduction. This was clearly demonstrated in Hartmann's experiments: for a period of 130 days, he removed daily one-third of the cytoplasm of an amoeba which normally divided every second day, so that the nucleus was left untouched. The amoeba did not die but neither did it divide, since it never reached the stage of growth necessary for cell division to occur.

The reproduction of multicellular organisms is only indirectly related to their growth, but growth is a direct consequence of the reproduction of their cells. In the case of plants it can be disputed whether their vegetative multiplication should be considered growth, multiplication, or regeneration.[G40] Therefore treating the abilities to grow and reproduce as a single criterion is rational practice.

2. A living system must have the capacity for hereditary change and, furthermore, for evolution,[S26] i.e. the property of producing increasingly complex and differentiated forms over a very long series of successive generations

Heredity is the capability of living systems to produce individuals identical or similar to themselves, or else to produce germ cells which ensure the development of such individuals. However, the criteria of reproduction include heredity, and so it would be redundant to consider heredity as a separate criterion.[S27]

However, the living world could not have evolved if heredity had followed strict rules, i.e. if the characteristics of the offspring had been always identical with those of the parents. In this case new characteristics could never have arisen in the living systems. Hence the capacity for hereditary change, i.e. the appearance of characteristics in the offspring which were not found in the endless chain of its progenitors, must form a separate criterion. This capacity is an indispensable condition for evolution, but it is not a prerequisite for the individual's living state. Hence it is included in the potential life criteria.

The capacity for hereditary change is a necessary but probably insufficient condition of evolution. A prerequisite of evolution is the possibility that hereditary changes with different (adaptive) 'values' can appear.[G41] This property can generally be found in the living world, but is not a criterion for individual life. Hence it is included in the potential life criteria.

3. Living systems must be mortal

Death is undoubtedly characteristic of living systems in the sense that non-living systems cannot die. However, this only means that death and life are connected to form a complementary pair of concepts. But death is indispensable to life in a deeper sense—death ensures the recycling of organic material. Without death exactly the same primaeval cells which consumed the organic substance of the biosphere at the beginning of life would still exist on Earth today.

Thus the death of the individual is indispensable to the living world. It is undoubtedly a life criterion, but it is not a criterion of the living state of the individual life, since the latter may cease without death. For example, in the case of cell division by binary fission and mitosis (because of the semiconservative replication mechanism of DNA) one of the two strands

[G40]Dawkins (1983, Chapter 14) considers the distinction between growth and reproduction in some provocative thought experiments. He identifies reproduction with passage through a single-cell 'bottleneck' which allows the developmental process to be 'reset' in each generation. This limits the concepts of reproduction and development to cellular and multicellular life and thus is inconsistent with Gánti's requirement that life criteria be valid on every level of the organizational hierarchy.

[S26]A living system cannot have the capacity of evolution; only a population of living systems has this capacity. Moreover, evolution does not necessarily lead to an increase in complexity. One should say that living systems may or may not be units of evolution *sensu* Maynard Smith (1986).

[S27]This statement is disputable. Heredity means that there are different *kinds* of multiplying entities, each reproducing its own kind. Heredity basically refers to the inheritance of *differences*. Haldane used to remark in his lectures that one could say that 'I inherited my nose from my father' and that 'I inherited my watch from my father'. Geneticists are concerned only with the first kind of inheritance.

[G41]This amounts to the requirement of heritable variance in fitness—the set of principles that Lewontin (1970) called 'Darwin's principles'. Lewontin's formulation of these principles is similar to, but differs in important ways from, Maynard Smith's (1987) view (see Griesemer (2000c) for a detailed comparison).

of the DNA molecules of the individual daughter cells comes from the mother, while the other is newly synthesized. The other cell substances are statistically distributed among the two daughter cells, which therefore contain in a statistically equal ratio the original substances of the mother cell and the newly synthesized substances. Thus the result is two cells which are equally young, neither of which can be considered a 'mother' or a 'daughter' cell. The life of these cells ceases during division in such a manner that no corpse is produced. Thus these cells are potentially immortal, at least in view of the continuity of their generation.

Cessation of the life of the individual without death is not restricted to unicellular organisms; it can also occur at the multicellular level. Simple multicellular animals (e.g. fresh-water polyps, planaria, and earthworms) can be cut into two or more pieces so that each part grows into a whole animal; thus, as in the case of cell division, the original individual ceases to exist but there is no corpse. This behaviour is well known in the vegetable kingdom (slip planting, vegetative multiplication).[S28]

[S28]This is why senescence appears only when (disposable) soma and (propagated) germ cells are distinguished. There is no senescence in the case of symmetrically dividing cells or multicellular organisms that can reproduce vegetatively at arbitrary positions of their bodies (e.g. hydra) (Kirkwood and Austad 2000).

Thus death is not an absolute but a potential criterion of life, which results from the existence of the living world.

All classical life phenomena (except movement as a simple change of place) are included in the above criteria but in a more exact and better-defined form. Furthermore, they include other criteria involving characteristics which have only been disclosed by molecular biology, such as the regulation and control of information processing and other processes. Hence the life criteria lifted above give a stricter prescription for the 'living' condition of a system than the classical life phenomena.

Since potential life criteria do not form a prerequisite of individual life, we consider a system to be living if it satisfies the real (absolute) life criteria, regardless of its concrete material construction or of the quality of the chemical substances forming it.

This definition provides a method of treating the fundamental laws of life completely generally, independent of their actual realizations. Thus this definition is not restricted to proteins, nucleic acids, or even to carbon compounds. If, in the future, exobiology (the science of life outside Earth) discovers non-carbon-based living systems, these will also be included on the basis of our general definition built upon the life criteria[S29] summarized below.

[S29]It is important that these criteria are separated from the currently known biota of our planet. How do we know that no casual criteria are included? The answer is simply that we cannot know this. But science rests on observation and manipulation of the real world. Therefore axiomatic constructions in science are prone to reconsideration if important new observations are made. This is why it is correct to call these life criteria rather than a definition of life.

- Real (absolute) life criteria

  1. Inherent unity
  2. Metabolism
  3. Inherent stability
  4. Information-carrying subsystem
  5. program control

- Potential life criteria

  1. Growth and reproduction
  2. Capability of hereditary change and evolution
  3. Mortality

3.6 Subsystems of the living cell

Starting from different points we reached the same conclusion in each of the previous sections, and we determined what to do in order to disclose the basic principle of life. Each case led to different requirements corresponding to the different paths taken. These requirements are not inconsistent; rather, they are complementary. Let us summarize them in order to see more clearly the work ahead of us.

First, we stated that only an exact theoretical biology could lead to a knowledge of the laws of life. Exact sciences are based upon abstract theoretical models; thus we have to construct the abstract model system of theoretical biology, which can be exactly described mathematically and which can approximate real biological phenomena with the required precision.

Then we stated that model systems of the exact sciences start from abstract basic units which clearly contain the qualitative properties characteristic of the disciplines, without any perturbations. Therefore we have to examine the abstract basic unit of biology which displays most clearly the characteristic qualitative property of biology, i.e. life.

Furthermore, we stated that systems displaying the fundamental biological property (i.e. life) can be found at different levels of an organizational hierarchy. If we are interested in the basic unit of life, we have to look at the lowest level where life is just appearing, i.e. on the level of prokaryotic cells.[G42]

We have also stated that present-day prokaryotes are highly complicated, perfected by a long period of evolution and functioning with the aid of highly complex enzymatic systems of regulation. But the nature of the basic units is not to be found in enzymatic regulation systems. Undoubtedly, these systems substantially increase the effectiveness of cellular functions and today no natural living system can exist without them, but the fundamental characteristics of life are carried in the system itself, which is regulated by enzymes. The basic principle of life must be sought in the organizational manner of the system. The organizational and functional laws of this regulated system will be discussed further below.

We have seen further that, in contrast with man-made dynamic technical systems, living systems are fundamentally soft, i.e. their processes are not carried out by devices composed of solid pieces in strict spatial connection, but run in solutions through spatially non-fixed chemical reactions although in an organized and regulated manner. The principle of life is to be sought in the organizational and functional laws of these soft systems.

A further restriction was imposed by the recognition that life is the property of unit systems, and so its basic principle must be sought in the way in which the soft unit systems are organized.

We have also seen that unit systems may have many accessory parts which may obscure the basic organizational characteristic of life. Thus we have to look for minimal systems which already display the characteristics of life, since the appearance of new properties is bound to the way in

[G42]A contemporary argument for this point is due to the theory of evolutionary transitions (Maynard Smith and Szathmáry 1995). If the levels of life evolved so that the members of higher levels are composed out of members of lower levels, then the basic units of life must include all the properties of living things without reference to any new properties that emerge as higher levels evolved.

which these systems are organized. Of course, this statement is equivalent to finding the principal basic units of biology. Hence the task is to construct an abstract and mathematically exact model of the minimal system which displays the characteristics of life.

We have stated that living systems are functioning dynamical systems which, despite the permanent changes that occur in them, exhibit remarkable internal stability. Processes leading to change are not consistent with a state of equilibrium; thus states of internal stability are non-equilibrium states. However, a system which is far from equilibrium can only be maintained at the price of performing continuous work. In living systems this continuous work occurs at the price of the energy of the nutrients, i.e. chemical energy. Thus the work done in living systems is chemical. Therefore we have to construct an abstract model of a soft minimal system which is capable of performing continuous chemical work.

A prerequisite of continuous work is that the changes in the system doing work should take place along 'constrained paths'. However, the elements of soft systems are spatially unrestricted. Hence the paths cannot be spatially constrained like those of mechanical, electric, and electronic devices. They must be of some other type, such as the changes in the state of the cooling gas in a refrigerator (although in living systems phase changes are unlikely to play an important role). Thus, in seeking the basic laws governing the functioning of living systems, we have to establish the organizational mode of the 'constrained paths' of soft chemical systems which are able to function continuously.

Nevertheless, this is not enough. Among all such systems, only those which satisfy the absolute life criteria are living. Even if we succeeded, at least theoretically, in constructing such a system, it would only represent a living individual. In order to be able to model the evolution of the living world, the system must also satisfy the potential life criteria.

Now that we have established the problem, we know exactly what to do. However, the task is not easy; at first glance it even seems to be insoluble. Indeed, it would be insoluble if we were unable to make use of the system–subsystem relation, which allows the task to be brokendown into simpler parts.

We have already discussed sets and systems in a separate section, but we have not considered subsystems. Now we have to understand the concept of subsystems. For simplicity, let us consider the example of a car. We have established that the characteristic qualitative properties of a system originate from the way in which it is organized (in this case its construction) and cannot be found in the properties of its elements. The properties of the car (it is a motor vehicle, it can run, the energy of motion is gained by burning petrol, etc.) cannot be found in the cogwheels of the gearshift, the valves of the engine, the steering device, the tyres, etc., but only in the system assembled from these. However, some of these qualitative properties are characteristic properties of the complete car, but individually are complete before the car is fully assembled. For example, the chassis is capable of travelling on the road though it cannot run on its own and one cannot sit in it. The engine produces energy by burning petrol but the car cannot run with it alone.

Parts of a complete system containing characteristic and necessary qualitative properties of the complete system are said to be subsystems of the complete system. Thus we can consider the chassis–steering device, the engine, the body the radiator, electric devices, etc. as subsystems of the car. The complete dynamic system is formed by the connected functioning of its subsystems. It contains the characteristic properties of its subsystems but it also has new qualitative properties. Thus, in the example of the car, none of its subsystems displays the property of a road vehicle.

Living systems can also be divided into subsystems. In man, for example, the circulatory system with the heart and blood vessels, the nervous system, the respiratory system, the excretory system, the hormonal system, etc. are subsystems which all have particular qualitative properties that are also characteristic of the whole system. Characteristic of man as a whole is that his heart beats, he thinks, he breathes, etc. However, the qualitative properties characteristic of man are only contained in the complete system organized from these subsystems. The complete system—man as a dynamic system—is constructed from its subsystems by the connected functioning of these subsystems; the function of each one is an indispensable condition for the functioning of the others. It is not the subsystem that is decisive here, but its function. Life is possible with a dialyser, a pacemaker, an artificial heart, or an iron lung, provided that the artificial subsystems carry out the function of the original subsystems.

Now we can not only see what to do but begin to see how to do it. We have to look for the subsystems of the simplest living systems and then construct abstract models of the soft minimal systems displaying the qualitative properties of the individual subsystems. Finally, we have to combine the subsystems into a single functionally uniform system. If all these tasks are performed well, we will obtain an abstract model of the simplest living system.

Three subsystems—the cytoplasm, the membrane, and the genetic substance—are present in every cell. There are many other subsystems with specific functions which are present in different types of cell, but these are not present in every cell. Therefore we can assume that if we establish the functions of these three subsystems, construct abstract models of them, and finally organize them into a single system, we will obtain an abstract model of the cell.[S30] Perhaps the cytoplasm has the most complicated function of the three subsystems. Firstly, it is the chemical motor. The cytoplasm contains the system transforming the chemical energy of nutrients into useful work. Secondly, it is the homeostatic subsystem compensating the influences of the external world by dynamically responding to them. Therefore the cytoplasm is responsible for the dynamic and organizational stability of the cell. However, it is also responsible for sensibility and excitability, since the accomplishment of homeostasis is nothing more than excitability. To achieve all these it is necessary that processes in the cytoplasm should occur in a regulated

[S30]This could be regarded as the most important insight of chemoton theory. The central tenet of the chemoton theory is that all three of these subsystems are needed in a minimal model of living systems. Keeping this in mind can help one navigate the complex problems of evolutionary transitions from one level of the organizational hierarchy to another, for example the origin of life from non-life, of cellular life, of multicellular life, and of socially organized life (Maynard Smith and Szathmáry 1995). For example, if an RNA world scenario for the origin of life is correct, then RNA molecules must have provided the functionality of all three subsystems. Similarly, if groups of socially organized organisms can themselves be said to 'live' as super-organisms, they must function as group metabolisms and membranes as well as 'genetic substances' in the sense of entities capable of 'group heritability'.

order, and thus the cytoplasm carries the property of regulation and is considered a soft system. Finally, the raw materials necessary for the growth and reproduction of all three subsystems (cytoplasm, cytomembrane, and genetic substance) are also delivered by the cytoplasm; thus it is a self-reproducing soft system. (This statement does not contradict the fact that the reproduction of the cytoplasm has preconditions ensured by the other two subsystems.)

The primary task of the cell membrane is to hold the cytoplasm together. Here we encounter a general requirement for the existence of soft systems which is not necessary in the case of hard systems. The basic condition for the existence of every system is that the distance between their elements is smaller than the maximal range of interactions organizing them into a single system. In hard systems this is accepted as a matter of fact: the cogwheels of the clock are *ab ovo* connected, the constituents of a radio are joined by wires, etc. Thus the structural organization of hard systems automatically ensures interactions of their elements. In the case of soft systems this does not always follow: a military division can only function when the orders can be sent to the soldiers, a herd can only function if the individuals belonging to it are within eyeshot and earshot, an ant-hill must be within the range of the action of its pheromones, and the cell could not function if its substances were dissolved in sufficient water to fill a swimming pool, since the necessary chemical interactions could not take place between molecules separated by such large distances.

Therefore appropriate spatial proximity of their elements is a necessary condition in soft systems. In human society this is ensured by mental activity, in a herd of animals by instinct, and in insect societies by pheromones. At the level of the cell and the organism, the spatial connection is ensured by membranes which are impermeable to their own internal substances. Thus the substances of the living system must occupy a definite spatial domain, ensuring the optimal concentrations necessary for the performance of chemical work and regulation.

However, the cell has metabolic activity and so it is not sufficient for the cell membrane to prevent the internal substances from diffusing into external space; it is essential that external substances needed as nutrients can move from the exterior of the cell into its internal space. Finally, metabolism produces a lot of waste products, and these must also pass through the cell membrane but in the opposite direction—from the interior of the cell to the external environment. Thus the task of the cell membrane as the second subsystem is on the one hand to limit the cell spatially and on the other hand to ensure appropriate connections with the external world for the first subsystem, the cytoplasm, to function.

The fundamental task of the third subsystem, the genetic substance, is to store information concerning the hereditary properties of the cell and to copy them during cell division. In same way as the regulation of the cellular processes is the task of the cytoplasm, the control of the life processes is the proper task of the genetic substance. The genetic material contains the program which, in the development of the cell, appears as a sequence of unidirectional irreversible events.

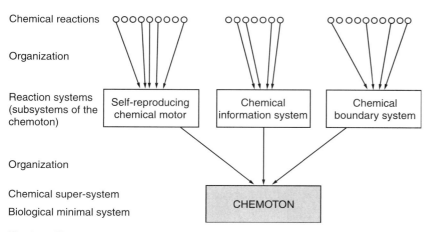

Chemical reactions

Organization

Reaction systems
(subsystems of the
chemoton)

Organization

Chemical super-system

Biological minimal system

Fig. 3.5 The organization of appropriate chemical reactions into a dynamic system results in reaction systems displaying new qualitative properties. By starting from these systems as elements or subsystems, one can obtain chemical super-systems at a higher level of the organizational hierarchy. One such system is the chemoton, which is also a biological minimal system.

Therefore, in seeking for the abstract minimal system of the cell, we have to construct abstract models of the following three subsystems.

1. A soft (chemical) system fulfilling the task of a chemical motor, i.e. of performing chemical work. This chemical motor must have a functionally stable inner organization, must be provided by chemical regulation, and must be capable of synthesizing chemical substances for itself as well as for other systems.
2. A soft (chemical) system which is capable of spatial separation, of being selectively permeable to chemical substances, and of growth in the presence of its raw materials.
3. A chemical system which is capable of storing and copying information, i.e. capable of self-reproduction in the presence of the appropriate raw materials.[G43]

In the following, we will show how to construct pure chemical systems corresponding to the functions of the individual subsystems, and then by uniting systems 1 and 2 as well as systems 1 and 3, we will obtain chemical supersystems displaying qualitatively new properties of a biological character. Finally, we shall see all the properties selected as life criteria appear in the chemical supersystem obtained by the union of the three systems—the chemoton (Fig. 3.5).

[G43]It is not obvious that the capacity of storing and copying information is the same as the capacity of self-reproduction in the presence of appropriate raw materials, although it may be the case that chemical systems capable of the one are (contingently) always capable of the other.

3.7 The chemical motor

Continuous work requires a work-performing system which changes its state during the work so that it always returns to its original starting state, i.e. it 'works' in a cycle. The internal combustion engine, the electric motor, the windmill, and the refrigerator have all been cited as examples.

A system which performs chemical work, where chemical energy is consumed, i.e. where it is continuously transformed by a chemical system, must also work in a cycle.

But what is a chemical cycle? It is certainly not a cyclic progression of individual molecules, which would in any case be impossible because of the random motion of molecules in an aqueous solution caused by thermal energy. In a chemical cycle the individual molecules are transformed by chemical reactions into other molecules which are transformed in their turn. After a number of transformations the original starting molecules are restored and the process can start again.[G44]

If molecules are denoted by A and qualitative differences by subscripts, the first step of the process is

$$A_1 \rightarrow A_2, \tag{3.1.1}$$

the second step is

$$A_2 \rightarrow A_3, \tag{3.2.2}$$

the third step is

$$A_3 \rightarrow A_4, \tag{3.3.3}$$

the penultimate step is

$$A_{n-1} \rightarrow A_n, \tag{3.3.n-1}$$

and the final step is

$$A_n \rightarrow A_1. \tag{3.3.n}$$

Thus, at the end of the process we have arrived back at our starting molecule (Fig. 3.6). If we wished to summarize mathematically what finally happened, it would seem to be nothing, since each of the terms occurs on both the left-hand side and the right-hand side of the equations, so that overall they are balanced.

In chemistry, however, the symbols mean material quantities; hence if A_1 can only be transformed into A_2, which can only be transformed into A_3, etc., eqns (3.1.1) to (3.1.n) mean that no chemical process resulting in a change of the sum total of the system's material can occur in this system, even if it 'functions', i.e. if chemical transformations are occurring in it.

This simple model serves two purposes simultaneously. First, it clearly reveals the meaning of 'constrained paths' in soft systems—they are chemical transformations in chemical systems such that these reactions can only occur through steps prescribed by the 'constraint'. For example, a given A_1 can only be transformed to A_2, which can only be transformed into A_3, which can only be transformed into A_4, and so on.[S31]

We can now see how to create dynamic stability in a system: it is only necessary to 'close' the 'constrained path' formed by the chemical

[G44]Rather, the original *kinds* of starting molecules are restored. The *same molecules* are not produced by the chemical reactions in the cycles any more than offspring bacteria are restorations of the original parent bacteria.

[S31]It is remarkable that in the same year as Gánti published this book, Otto Rössler published a short paper arguing for the central role of such catalytic cycles in biogenesis, emphasizing a system-theoretic viewpoint (Rössler 1971).

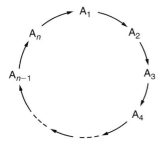

Fig. 3.6 In a chemical cycle the molecules are transformed in every step but the original molecules are restored at the end of the cycle. This type of cycle also satisfies Ashby's cybernetic stability criterion.

reactions, i.e. to bring it back to the starting point via a chemical cycle. In the example we have to ensure that A_1 is re-transformed to A_1, such that the cycle is ready to start again. Thus the system will be inherently stable, i.e. stable by its very nature, since a cycle is always re-formed into itself. This also satisfies Ashby's 'cybernetic' stability criterion, which requires that 'the sequence of transformations should not create new states'. (Of course, inherent stability is not identical with thermodynamic stability.) Inherent stability is not yet the stability of growing accumulation systems but, as we shall see later, it is already its basis (Gánti 1971). It can be proved that inherent stability is valid not only for simple reaction cycles but also for the most complicated closed network of chemical constrained paths.

The first biochemical cycle was identified by Hans Krebs on the basis of fundamental discoveries made by Albert Szent-Györgyi. This cycle is now known as the Szent-Györgyi–Krebs cycle, the citric acid cycle, or the tricarboxylic acid cycle. In this cycle organic acids are transformed into each other:

$$\text{oxaloacetic acid} \rightarrow \text{citric acid} \qquad (3.2.1)$$

$$\text{citric acid} \rightarrow \text{isocitric acid} \qquad (3.2.2)$$

$$\text{isocitric acid} \rightarrow \text{ketoglutaric acid} \qquad (3.2.3)$$

$$\text{ketoglutaric acid} \rightarrow \text{succinic acid} \qquad (3.2.4)$$

$$\text{succinic acid} \rightarrow \text{fumaric acid} \qquad (3.2.5)$$

$$\text{fumaric acid} \rightarrow \text{malic acid} \qquad (3.2.6)$$

$$\text{malic acid} \rightarrow \text{oxaloacetic acid.} \qquad (3.2.7)$$

It can be seen that in this cycle oxaloacetic acid is transformed through a series of chemical reactions back into oxaloacetic acid, i.e. despite continuous chemical transformation 'nothing happens' (Fig. 3.7). However, if nothing really happens, what is the aim, reason, role, or task of this cycle? If nothing is really happening here, is not this cycle a kind of *perpetuum mobile*, permanently cycling without any energy source?

However, these are the wrong questions since they consider only the compounds constrained to 'cyclic paths'. Chemical reactions (apart from some intramolecular transformations) take place via the interaction of two different compounds, or at least produce two reaction products. Thus A_1 needs a reactant X_1 for its transformation or another reaction product Y_1 is formed in addition to A_2, which, however, will not be used in its free state but bound to hydrogen-bearing compounds:

$$A_1 + X_1 \rightarrow A_2 \qquad (3.3.1)$$

or

$$A_1 \rightarrow A_2 + Y_1 \qquad (3.3.2)$$

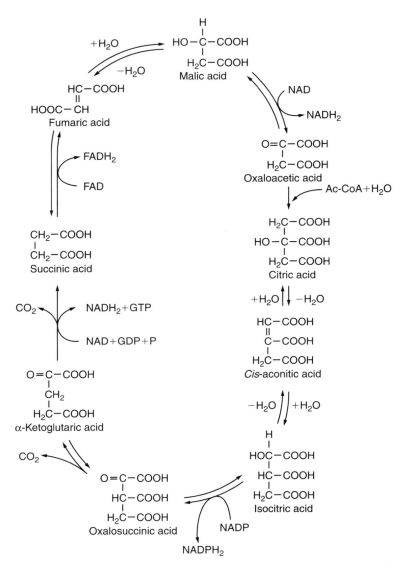

Fig. 3.7 The reaction system of the Szent-Györgyi–Krebs (citric acid) cycle. The cycle 'consumes' acetyl groups; their 'burning' into carbon dioxide yields the energy necessary for the cycle to function and for the performance of work. As a result of work, hydrogen is produced which is not used in its free state but is bound to hydrogen-bearing compounds. Ac–CoA, acetyl CoA.

or

$$A_1 + X_1 \rightarrow A_2 + Y_1 \tag{3.3.3}$$

The same is true for the other reaction steps (Fig. 3.8).

Now, if we examine the Krebs cycle from this point of view, any possibility of a *perpetuum mobile* disappears. In the Krebs cycle carbon is

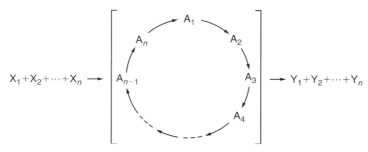

Fig. 3.8   A chemical cycle can only work at the price of external energy, mostly the chemical energy of the nutrients $(X_1 + X_2 + \cdots + X_n)$ that are introduced. The total energy of the waste products $(Y_1 + Y_2 + \cdots + Y_n)$ is always less than that of the nutrients consumed.

'burned' into carbon dioxide and the energy released drives the cycle. In the first step of the cycle a structure with two carbon atoms, the acetyl group CH_3–CO– is bound to the oxaloacetic acid, and the oxaloacetic acid molecule which contains four carbon atoms becomes a citric acid molecule with six carbon atoms:

$$\begin{array}{ccccc}
\text{C–C} & + & \text{C–C–C–C} & = & \text{C–C–C–C–C–C} \\
\text{(carbon skeleton} & & \text{(carbon skeleton of} & & \text{(carbon skeleton} \\
\text{of acetyl group)} & & \text{oxaloacetic acid)} & & \text{of citric acid)}
\end{array}$$

In subsequent transformations two carbon dioxide molecules are removed from the citric acid, leaving products with four carbon atoms. In the final step oxaloacetic acid is produced and is ready to start the cycle again. Ignoring chemical details, let us introduce the notation

$$\text{oxaloacetic acid} \xrightarrow[\;\;①\;\;]{\text{Krebs}} \text{oxaloacetic acid} \qquad (3.4)$$

to represent a single passage of the oxaloacetic molecule through the Krebs cycle—a constrained path. Using this notation, we can write

$$\text{oxaloacetic acid} + CH_3\text{–CO–} \xrightarrow[\;\;①\;\;]{\text{Krebs}} \text{oxaloacetic acid} + 2CO_2.$$

$$(3.5)$$

Thus it can be seen that two carbon atoms are 'burned' to make carbon dioxide, while an oxaloacetic acid molecule 'goes once around' the cycle. However, this 'burning' only releases just enough energy to make the cycle function.

Clearly, this is not a *perpetuum mobile*; the chemical cycle is driven by the chemical energy of the external substance (in this case the acetyl group). Therefore the chemical energy content of the resulting by-products is less than that of the reactants—the raw materials. This motor can only work if a raw material with a higher energy content, i.e. the external reacting substance, is present. Thus we have answered one of our questions. The other one remains to be answered: What use is a chemical motor like this?

At a general level we have already provided an answer in one of the preceding sections—it can be used for work-performance, for doing chemical work, for performing useful directed work. Now we have to establish the mechanism of this work.

First let us consider an example of chemical work-performance the production of hydrogen as a fuel for rockets. Hydrogen is used as a rocket fuel because its burning to water releases a large amount of energy and the efficiency of the energy gain relative to weight is very good. Hydrogen is produced industrially on a large scale by breaking down water using energy provided by electricity. Hydrogen can also be obtained from water by chemical energy, but electrical energy is more efficient under industrial conditions.

Interestingly, living organisms discovered that he burning of hydrogen to water is an efficient energy source thousands of millions of years ago; the energy supply of aerobic (air-using) organisms is based on this process. However, living organisms do not burn hydrogen in its gaseous form, but use a special form of chemically bound hydrogen. Similarly, the combustion does not take place at a high temperature, but through a series of complicated chemical reactions so that the minimum amount of energy released will be wasted as heat. But how does the organism obtain combustible hydrogen?

Readers with a knowledge of chemistry will have noticed that eqn (3.5) incorrect. True, there is enough carbon for the production of two carbon dioxide molecules, but there are three less oxygen atoms on the left-hand side of the equation than on the right-hand side, and three hydrogen atoms are missing on the right-hand side. This provides the solution. Namely, there are three steps in the cycle in which the external reacting substance is water. The oxygen from these water molecules is used to form carbon dioxide and the remaining hydrogen atoms are bound to hydrogen-bearing compounds in a form ready to be 'burned':

$$\text{oxaloacetic acid} + CH_3-CO- + 3H_2O$$

$$\xrightarrow[\;\;\;\;\;\;\;\;]{\text{Krebs}} \text{oxaloacetic acid} + 2CO_2 + 9H. \tag{3.6}$$

Therefore the citric acid cycle produces hydrogen by breaking down water, and to do this it uses the energy released in the oxidation of carbon to carbon dioxide! Therefore it is indeed a soft system suitable for performing continuous work—a system capable of directed useful work by making use of chemical energy! (In order to avoid misunderstandings, let us note than eqn (3.6) is still oversimplified and incomplete.)

We still have to determine whether this system is really capable of continuous performance of work. According to eqn (3.6) a single oxaloacetic acid molecule makes two carbon dioxide molecules and nine hydrogen atoms from one acetyl group and three water molecules, if all chemical transformations of the cycle are performed just once. At the end of the cycle the oxaloacetic acid molecule which has been restored can go through the cycle again, so that a total of four carbon dioxide molecules and eighteen hydrogen atoms are produced. This can be expressed in an equation as follows:

$$\text{oxaloacetic acid} + 2(\text{CH}_3\text{-CO-}) + 6\text{H}_2\text{O}$$

$$\xrightarrow[\text{Krebs}]{\textcircled{2}} \text{oxaloacetic acid} + 4\text{CO}_2 + 18\text{H}. \qquad (3.7)$$

This process could be repeated a hundred times, giving the result

$$\text{oxaloacetic acid} + 100(\text{CH}_3\text{-CO-}) + 300\text{H}_2\text{O}$$

$$\xrightarrow[\text{Krebs}]{\textcircled{100}} \text{oxaloacetic acid} + 200\text{O}_2 + 900\text{H}. \qquad (3.8)$$

Indeed, the process can be continued until there are no $\text{CH}_3\text{-CO-}$ or water molecules present. Therefore the chemical cycle is capable of continuous directed useful performance of chemical work.

It is quite reasonable to generalize the above statements, since many cycles exist and all have characteristic properties in common. However, a relevant generalization requires a general formula. Now I must ask the reader to follow closely the simple calculations below which will not go beyond simple arithmetic—mostly addition with occasional multiplication. (I have to mention this in advance, since the chemoton theory has been objected to as being just complicated mathematics without any connection to reality, and as being empty biological gobbledygook without any relevant mathematical formulae, such as set theory.) Some minimal formalism is necessary to understand chemoton theory, but the basis of this formalism is just the general description of cyclic processes.

Let the internal components of a cycle (in the case of the Krebs cycle the organic acids) be denoted by $A_1, A_2, A_3, \ldots, A_n$. Thus in the case of the citric acid cycle oxaloacetic acid is denoted by A_1, citric acid by A_2, and so on. The reacting substances necessary for the individual steps are denoted by $X_1, X_2, X_3, \ldots, X_n$, and the resulting reaction products by $Y_1, Y_2, Y_3, \ldots, Y_n$. We have seen that the last two types of substances did not cause much trouble, so let us represent them by single symbols:

$$X_1 + X_2 + X_3 + \cdots + X_n = \sum X_i \quad \text{or} \quad \text{simply } \mathbf{X}$$

and

$$Y_1 + Y_2 + Y_3 + \cdots + Y_n = \sum Y_i \quad \text{or} \quad \text{simply } \mathbf{Y}.$$

In the following, **X** simply means that every reacting substance is present in the necessary amount and **Y** means that all reaction products arise in amounts corresponding to the reaction equation. Thus eqn (3.6) describing the functioning of the Krebs cycle can be written as follows:

$$
\underset{\substack{\text{(oxaloacetic}\\\text{acid)}}}{A_1} \quad + \quad \underset{\substack{\text{(reacting}\\\text{substances)}}}{\mathbf{X}} \quad \xrightarrow{\text{Krebs}} \quad \underset{\substack{\text{(oxaloacetic}\\\text{acid)}}}{A_1} \quad + \quad \underset{\substack{\text{reaction}\\\text{products}}}{\mathbf{Y}.}
$$

$$(3.9)$$

This equation can be also interpreted as a summary of the following process: if one oxaloacetic acid molecule in the presence of all the necessary reacting substances goes through every transformation prescribed by the Krebs cycle once, it will be restored as oxaloacetic acid, while every one of the necessary reacting substances will give one or more other molecules prescribed by the chemical reaction equations. So far so good, since we have not done anything more than substitute the individual compounds in the Krebs cycle by letters.

However, a circle is characterized by having no end and no beginning; it can be started anywhere. The same holds for a cyclic process. If a citric acid molecule (A_2) is put into the reaction mixture in place of an oxaloacetic acid molecule and is allowed to pass through all the steps of the Krebs cycle until it becomes citric acid again, the same reacting substances will be used in the same amounts and the same reaction products will be obtained in the same amounts as if an oxaloacetic acid molecule had gone through the reactions of the cycle. Therefore we have

$$
A_2 + X \xrightarrow{\text{Krebs}} A_2 + Y \qquad (3.10)
$$

and, by the same principle,

$$
A_3 + X \xrightarrow{\text{Krebs}} A_3 + Y \qquad (3.11)
$$

or

$$
A_4 + X \xrightarrow{\text{Krebs}} A_4 + Y \qquad (3.12)
$$

and so on. A molecule of any one of the components of a cycle requires the same reacting substances and produces the same reaction products.

In chemistry, one does not work with a single molecule of a substance, but with a large number of them. Thus eqns (3.9), (3.10), (3.11), etc. can be taken to refer, by definition, to molar quantities (i.e. to 6×10^{23} molecules) also. However, in the case of many molecules, nothing ensures that only component A_1 or only component A_2, etc. is present in the system. We have to consider their mixture:

$$A_1 + A_2 + A_3 + \cdots + A_n = \sum A_l \quad \text{or} \quad \text{simply } \mathbf{A}.$$

Now each of the components, in going once through the cycle, consumes and produces the same substances. Therefore the individual components can be replaced by their mixture:

$$\mathbf{A} + \mathbf{X} \xrightarrow{\text{Krebs}} \mathbf{A} + \mathbf{Y}. \tag{3.13}$$

This is an important result, since this equation states that in any mixture of the acids of the Krebs cycle, the same products are obtained in the same amounts if each of the molecules of the mixture goes once through the reactions of the cycle.

Finally, this result can be generalized. In any mixture of the components $A_1, A_2, \ldots, A_n$ of a chemical cycle A, the same reaction components are produced in the same amounts if each of them takes part once in every one of the reactions of cycle A:

$$\mathbf{A} + \mathbf{X} \xrightarrow{\text{Krebs}} \mathbf{A} + \mathbf{Y}. \tag{3.14}$$

I can imagine that some biologically oriented readers may now heave a sigh: what a superfluous and senseless complication this is! At the same time, chemically oriented readers may consider eqn (3.14) to be a primitive and even incorrect oversimplification! In a certain sense, the latter are right, but the exact stoichiometric description is far too complicated and can easily be found in the literature. However, I have to tell the biologists that, although it appears to be a 'superfluous' complication, this simple equation is the basis of a theory leading directly to biologically interesting results.

Let us now examine eqn (3.14). It is easy to see that **A**, which occurs on both sides of the equation, could be neglected from a mathematical point of view. Indeed, chemists did not bother themselves much with it, and compounds like A were simply neglected in their stoichiometric equations. They realized, of course, that substances which help and accelerate certain chemical transformations without any change in their quantity were present, but they did not realize that these substances function in cycles. Therefore they simply neglected these 'catalysts' in their equations, or at most indicated that they were necessary by some sign above the arrow representing the reaction. Chemists were mainly interested in the compounds undergoing transformation.

However, the main interest of biologists is not the transformation of one thing into another, since they know that animals transform organic compounds into carbon dioxide and water, yeasts transform sugar into alcohol and carbon dioxide, and plants transform carbon dioxide and water into organic compounds and oxygen. Biologists are primarily interested in the transforming agent producing the changes, since without this the process will not take place at all. This transforming agent (except in the case of growth which is dealt with in the next section) is present on both sides of the equation and was denoted above by **A**. In the simplest case **A** represents a chemical cycle; in the most complicated case it represents the complete metabolic network of a highly developed organism, say man.

Berzelius considered yeasts not as organisms but as simple catalysts which aided the transformation of glucose into alcohol. Of course, he was not aware of the highly complicated functioning of yeasts. However, his mistake was not too large: the basic principles of both simple catalysis and the complicated metabolism of yeasts are cyclic processes. These cyclic processes appear on both sides of the fundamental equations.

Equation (3.14) also contains the principle of inherent stability: **A** does not change during the course of the chemical processes. Since **A** is unchanged, eqn (3.14) shows that the driving force of the process is in the **X → Y** transformation, i.e. in the raw material or 'nutrients'. From a thermodynamic point of view the feasibility of the process is the condition that the energy content of the sum of the raw materials should be larger than that of the sum of the products:

$$E_X > E_Y.$$

(or, more exactly, the free enthalpy of the sum of the raw materials should be greater than that of the sum of the products: $G_X > G_Y$). Therefore the 'chemical motor' only works if the raw material can transform from a higher to a lower chemical energy level, just as in the case of a water wheel where water has to flow from a location with higher potential energy to one with lower potential energy, or an electric motor where current must flow from a higher to a lower electric potential.

Biologists must primarily be concerned with this chemical motor, since this system of chemical cycles is the basis of the functioning of life (Gánti 1971). This chemical motor lies along the downward path of chemical energy in order to turn the gradient into work, just as the waterwheel intercepts the fall of the potential energy of water, the electric motor intercepts the current flowing from a higher to a lower electric potential, and the windmill turns in the direction of the decrease of atmospheric pressure—all in order to convert that part of the released energy 'caught' by these motors into useful work (Fig. 3.9).

Finally, biologists have to realize from eqn (3.14) that the condition $G_X > G_Y$, i.e. the requirement that the sum of the energy content (free enthalpy content) of the nutrients should be greater than the energy content of the products, although being a necessary prerequisite for the chemical motor to function, does not preclude the energy content of individual products from being greater than that of any of the nutrients, provided that this excess is compensated by a lower energy content in the other products. The real importance of this consequence will be seen in the next section.

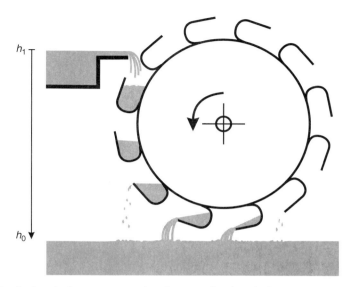

Fig. 3.9 A chemical motor uses the decrease in chemical energy in the same manner as the water-wheel uses the fall in the potential energy of water. In both cases, part of the energy will be transformed into useful work.

3.8 Growing systems

It is tempting to compare the properties of our chemical motor with those of the cytoplasm, which is one of the subsystems of the cell. Obviously, they have some properties in common: both are functioning systems, both are soft systems, both function by consuming chemical energy, both are inherently stable, both are capable of transforming 'nutrients' into other kinds of chemical substances, both can continuously produce new compounds, etc. However, they also differ in many respects, and the most important of these differences is the cumulative property of the cytoplasm: the cytoplasm is a system which is capable of accumulation (i.e. growing) but the chemical cycle is not.

Crystals also grow in saturated solutions. However, this is not an active but a passive growth, which is entirely different from the growth of the dynamic systems called accumulation systems. The growth of these systems does not require a saturated solution; they can often take up the substances necessary for their increase from highly dilute solutions. Moreover, they do not obtain their building substances in ready-made form; in contrast, they make them themselves from other substances by means of chemical reactions.

Now, substances which can promote their own synthesis by chemical reactions are known. These substances are called autocatalysts, since they catalyse their own synthesis, i.e. their own production. From this point the solution is a single step. We have seen that catalysts function in cycles, i.e. a catalyst (more exactly a homogeneous catalyst) is not a single substance—a single compound—but a cyclic system with components functioning in the same manner. Thus autocatalysis is not the function of a single substance—the autocatalyst—but is the result of an autocatalytic chemical cycle, i.e. a chemical reaction system.

The basic principle of the mechanism is quite simple. We have seen that a chemical cycle performs directed chemical work in the sense that it produces chemical compounds, i.e. it synthesizes a variety of chemical substances. We introduced a single general requirement for these synthesized substances: the total energy (free enthalpy) content of the reaction products had to be smaller than that of the source materials. Provided that this condition is satisfied, freedom is complete, and the properties of the given cycle determine what kind of chemical substances can be produced by it. For instance, the Krebs cycle produces hydrogen (or more exactly hydrogenated hydrogen carriers). Other cycles produce other types of compound. Why should a cycle not synthesize one of its own internal components? Indeed, such cycles have been discovered in the last two decades, although autocatalytic character was not recognized. It was pointed out by the author in 1971 that the malic acid cycle, the Calvin cycle, and the phosphorylation of adenine to adenosine triphosphate (ATP) are all autocatalytic cycles. Since the idea of autocatalysis in chemistry, and in particular in biochemistry, is not always unambiguous, in what follows we shall call these autocatalytic cycles 'self-reproducing cycles', particularly because this name is much more descriptive of the specific properties of these chemical systems.

The Krebs cycle could only be a self-reproducing cycle if an oxaloacetic acid molecule passing through the constrained path of the reactions, finally led to two oxaloacetic acid molecules:

$$A_1 + X \xrightarrow{\text{Krebs}} 2A_1 + Y. \tag{3.15}$$

However, there is such a cycle, namely the malic acid cycle or malate cycle (Fig. 3.10). Its components are almost identical with those of the Krebs cycle, except that the 'constrained paths', the source materials, and of course the end-products differ. In the malate cycle the acetyl groups built in do not burn to form carbon dioxide. Instead, two acetyl groups are synthesized to malic acid, which contains four carbon atoms, and this in turn is transformed into oxaloacetic acid:

$$A_1 + X \xrightarrow{\text{Malate}} 2A_1 + Y. \tag{3.16}$$

Thus the starting oxaloacetic acid is duplicated during each cycle. Under continuous functioning the next cycle starts with twice as much oxaloacetic acid:

$$2A_1 + 2X \xrightarrow{\text{Malate}} 4A_1 + 2Y. \tag{3.17}$$

Thus at the end of the second cycle the original quantity of oxaloacetic acid is quadrupled. It is easy to see that at the end of the third cycle the

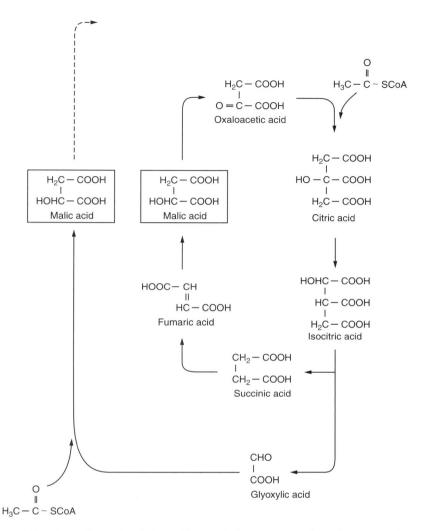

Fig. 3.10 The malic acid cycle is a self-reproducing autocatalytic cycle. The starting amount of malic acid is doubled in every cycle, and the system consumes acetyl-coenzyme A.

quantity of oxaloacetate present is eight times the original, at the end of the fourth 16 times, then 32 times, 64 times, etc. In other words, the amount of starting material increases by the power of 2 as a function of the number of cycles.

Every member of the cycle functions in the same way and thus an equation corresponding to eqn (3.16) can be written for any of its elements. For example, for citric acid (A_2)

$$A_2 + X \xrightarrow[\enspace \textcircled{1} \enspace]{\text{Malate}} 2A_2 + Y \qquad (3.18)$$

and for malic acid (A_7)

$$A_7 + X \xrightarrow[\text{①}]{\text{Malate}} 2A_7 + Y. \tag{3.19}$$

Just as in the case of simple chemical cycles, it follows from this that the functioning of a self-reproducing cycle does not depend on the ratios of the components present, and thus it can be written

$$\mathbf{A} + \mathbf{X} \xrightarrow[\text{①}]{\text{Malate}} 2\mathbf{A} + \mathbf{Y}, \tag{3.20}$$

i.e. we have to deal with the synthesis of the complete self-reproducing cyclic system and not with that of a single substance!

Not all self-reproducing cycles are so simple. Some are very complicated.

One example is the Calvin cycle (Fig. 3.11), which incorporates carbon dioxide into plant tissue during photosynthesis, i.e. synthesizes of carbohydrates from carbon dioxide. This process was discovered by the American biochemist, Melvin Calvin, who was consequently awarded the Nobel Prize.

In the Krebs cycle organic compounds are 'burned' to carbon dioxide whilst energy is released. Conversely, if we wish to synthesize organic compounds from carbon dioxide, we have to use chemical energy. Plants obtain their energy requirement from light energy, and this must first be transformed into chemical energy. This is done by breaking up water: oxygen diffuses away into the air and hydrogen becomes bound to a hydrogen-bearing compound ($NAD + H_2 \rightarrow NADH_2$). Another compound appropriate for energy storage is also synthesized (ATP). This part of photosynthesis is called the light reaction, since it only occurs in the presence of light.

In the second period of photosynthesis the energy-rich compounds ($NADH_2$, ATP) thus obtained are used in the Calvin cycle to synthesize carbohydrates (more exactly, sugar phosphates). Now, the only source materials required for this synthesis are carbon dioxide and water, although the system itself is very complicated, consisting of a reaction network organized from 28 chemical reactions.

Thus carbon is incorporated into the living world by a self-reproducing cycle utilizing organic carbon compounds, just as phosphorus is by auto-catalytic cycles utilizing organic phosphates. Nitrogen also appears to enter living organisms through a system of self-reproducing cycles involving organic nitrogen compounds. Thus self-reproducing cyclic systems can be found in the fundamental growth processes in the real living world (Gánti 1971).

We are now able to compare the functions performed by the cytoplasm with the properties of self-reproducing reaction networks. We have noted earlier that the cytoplasm is the chemical motor transforming the energy of the source materials into useful work. This is exactly what is done by all chemical cycles, including the self-reproducing cycles. We have seen that

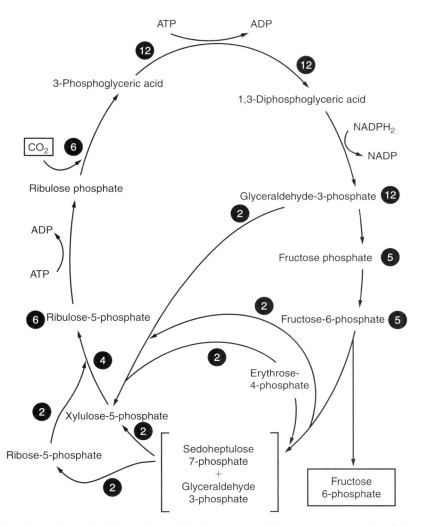

Fig. 3.11 Simplified illustration of the Calvin cycle. The inner compounds of the cycle are sugar phosphates which transform into each other and the 'nutrients' is carbon dioxide, bound hydrogen and bound phosphate. The cycle gains the energy necessary for the transformation of carbon dioxide into sugar phosphates, i.e. into its own inner compounds, from the bound hydrogen and phosphate.

the processes of the cytoplasm are regulated processes. This is also true for every chemical cycle, since a cycle is, by definition, a feedback process, which thus satisfies the basic requirement of regulation. (We shall return to this question in one of the following sections.)

Further, we have explained that the cytoplasm is the homeostatic subsystem of the cell, carrying the property of inherent stability and being responsible for excitability. We have already seen that, by definition, all simple chemical cycles are stable, since the only changes which can occur in a cycle are those which always lead to the same states. But this inherent stability was not identified with the stability of the cytoplasm, since the latter is also stable during growth while the simple chemical cycles are not growing systems.

The process of self-reproduction does not affect the property of stability. This can easily be seen in the example of the malic acid cycle. Let us begin with a single malic acid molecule and start the process. The malic acid is transformed into citric acid, which is transformed into isocitric acid, etc. until finally malic acid is formed again. In fact, two molecules of malic acid are formed, but the second malic acid molecule does not prevent the first molecule from going through the 'constrained paths' of the cycle again if there are enough source materials present and the solution is of the appropriate strength. The mechanism for regulating increase in malic acid is very simple: if the glyoxylic acid formed in the cycle is reacts not with acetyl-coenzyme A but with some other molecule, the self-reproducing character of the process ceases but the 'chemical motor' continues to run, i.e. regulating the increase does not affect the functioning of the chemical motor in this system.

At the same time the 'chemical motor' detects external changes: every change in the thermodynamic parameters of the environment (temperature, pressure, concentration, pH, etc.) affects upon the rate of the individual chemical reactions. Thus the function of the whole system depends on this effect; the system performs the 'chemical work' more or less intensively without any change in its structure. Thus it seems that self-reproducing cyclic processes also satisfy the condition of homeostasis.

Finally, we have seen that the cytoplasm delivers the source materials which enable it to reproduce itself and the two other subsystems. The self-reproducing cycles can also produce the substances necessary for their own reproduction; we shall see in the following sections that they can also do this for the other two subsystems.

In view of the many properties that the functioning of cytoplasm and of self-reproducing chemical cycles have in common, it is clear that we can choose, the latter as a subsystem of the abstract minimal system to represent the basic unit of life. However, there are many kinds of self-reproducing cycles; obviously, we must choose the one which is theoretically simplest and which represents the basic properties of interest to us in the simplest form. Clearly, this will be a cycle with three steps: the first step is the reaction with the 'source material' X, the second step is the production of the by-product Y, and the third is self-reproduction:

$$A_1 + X \rightleftharpoons A_2 \tag{3.21}$$

$$A_2 \rightleftharpoons A_3 + Y \tag{3.22}$$

$$A_3 \rightleftharpoons 2A_1. \tag{3.23}$$

Then by summing and simplifying we obtain

$$\textbf{A} + \textbf{Y} \xrightarrow[\;\;\;1\;\;\;]{A} 2\textbf{A} + \textbf{Y} \qquad (3.24)$$

which is shown graphically in Fig. 3.12.

3.9   Chemical systems multiplying by division

A farmer who wants rapid results will sow annual plants and in a few months will be able to harvest the crop. However, the next year he has to start again. If he plants a fruit-tree, he has to work without any gain for several years, but once the tree begins to bear fruit, he can take profit from his work every year for decades.

Something like this has happened to us. We have invested a great deal of work in the last eight sections, and now it is time for the results. They will indeed come, the first this section, and later we shall be able to obtain new results section by sections, with only a little more work. However, this additional work cannot be neglected. In the present section we shall have to find a subsystem which can be connected with the subsystem introduced in the previous section so that we will be able to profit from our work. For simplicity, we shall look for a system which can satisfy the role of the cell membrane.

We shall not need to look very hard; the system has already been provided by nature. Cells are surrounded by a membrane of electron-microscopic thickness, and similar membranes also enclose the cellular components, even in the cytoplasm itself and within mitochondria and chloroplasts. For a long time we did not understand membranes and we had no model with which to interpret their properties. Then, in 1972, Singer and Nicholson introduced the two-dimensional liquid membrane model. Therefore we only need to generalize the statements about this model and write down its equation.

Some compounds, such as salts and sugars dissolve easily in water but not in alcohols, benzene, or ethers. Other compounds, such as fats and oils, dissolve easily in alcohols, ethers, and similar organic solvents, but are insoluble in water. Solubility can be explained in terms of molecular properties: if the molecule contains hydrophylic groups, it is soluble in water; if it contains hydrophobic groups, it will easily dissolve in organic solvents (the lipid solvents).

However, what happens if a long rod-like molecule has a hydrophylic group at one end and a hydrophobic group at the other end? In a system consisting of a benzene layer on water will it dissolve in the benzene or the water? Such molecules arrange themselves on the border separating water from benzene with their hydrophobic part in the benzene phase and their hydrophilic part in the water phase. Thus they form a monomolecular layer on the benzene–water interface (Fig. 3.13). Molecules can be removed from this layer into either the water or the benzene phase only if energy is put into the system. Conversely, spontaneous incorporation of

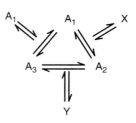

Fig. 3.12 Elementary abstract model of the soft minimal system corresponding to the cytoplasm. Thus we have found one of the subsystems of the basic unit of theoretical biology.

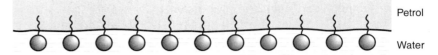

Petrol

Water

Fig. 3.13 Rod-like molecules with hydrophilic groups at one end and hydrophobic groups at the other are situated at the interface of water and benzene with their hydrophobic groups in the benzene and their hydrophilic group in the water.

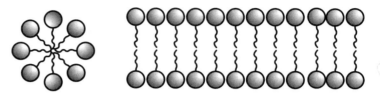

Fig. 3.14 If these molecules are forced into water, they will arrange themselves so that only their hydrophilic are in contact with water. This ordering may occur in the form of molecular aggregates or double layers.

molecules into the monomolecular boundary layer from either the water or the benzene phase is accompanied by an energy release.

If molecules with this double nature are forced into the aqueous phase, they arrange themselves with the hydrophobic groups facing each other, thus forming a local water-free environment. This may occur in the form of either aggregates or double molecular layers. In such circumstances the hydrophilic groups of the molecules on the surface of the aggregate or the membrane face outward into the water, while the hydrophobic groups face inward and produce conditions like those that would exist they were in an organic solvent (Fig. 3.14).

These membranes have interesting properties. Molecules can only be removed from them by investing energy; therefore in this respect membranes are like solids. However, in the plane of the membrane, the molecules are not bound by anything; they can easily move in any direction in the plane and they can replace each other, i.e. the membrane behaves like a fluid. It was for this reason that Singer and Nicholson called these membranes two-dimensional liquids (Fig. 3.15).

Living systems are always surrounded by membranes based on double molecular layers, even if they are additionally covered by cell walls, calcareous shells, or chitin cuticula. The membranes within the cells are also formed from double molecular layers. The basic components of natural two-dimensional liquid membranes are phospholipids, but membranes can also be produced artificially from many compounds, provided that the molecules have hydrophobic groups at one end and hydrophilic groups at the other.

As can be seen in Fig. 3.15, the hydrophobic groups at the edge of the membrane are still in contact with water. Therefore the edge of the membrane is an unstable system which is trying to close and form a spherical surface. If membrane formation is initiated in an aqueous

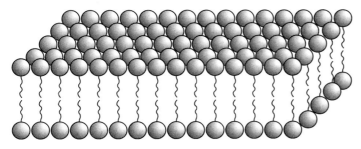

Fig. 3.15 A molecule can only be removed from a double-layer membrane by performing work. Thus the membrane behaves like a solid body. However, the molecules can move freely within the plane of the membrane. Thus they behave like fluids. For this reason, Singer and Nicholson called such membranes 'two-dimensional fluids'.

solution, the result will be many microscopic spherules bounded by membranes with a double molecular layer.

These membranes are also capable of growth. This growth is similar to that of crystals, since the membrane-forming molecules in the solution are incorporated into the membrane spontaneously by physical effects and chemical processes are involved. This corresponds to the statement above that membranes behave like solids with respect to liquids. Membrane growth can also be initiated on either of its two sides, and occurs on both sides since molecules can jump from one layer the other.

We now have to describe the membrane growth process quantitatively, using an equation. Yet this is a very simple task indeed. Let the membrane-forming molecules be denoted by T and the membrane built from n molecules T by $\boxed{T_n}$. The incorporation of one molecule T into the membrane can be described by

$$\boxed{T_n} + T \rightarrow \boxed{T_{n+1}}. \tag{3.25}$$

Similarly, if n molecules T are incorporated into the membrane $\boxed{T_n}$ we have

$$\boxed{T_n} + nT \rightarrow \boxed{T_{2n}}. \tag{3.26}$$

However, the membrane becomes larger with the addition of every molecule, and if the number of molecules forming the membrane doubled, the surface of the membrane is also reduplicated. Thus $\boxed{T_{2n}}$ also represents a surface that is twice the size of surface $\boxed{T_n}$.

Let us now construct the abstract model of a chemical system dividing in space by connecting the function of the autocatalytic cycle to membrane synthesis. The simplest form of connection is the requirement that the membrane-forming molecules should be produced by the auto-catalytic cycle; therefore the membrane can only grow if the autocatalytic cycle is functioning. In order to do this our autocatalytic minimal system must consist of four steps: we must add to the three steps discussed in the previous section a fourth reaction producing the basic substance of the

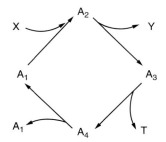

Fig. 3.16 The self-reproducing minimal cycle of a proliferating chemical supersystem. In addition to producing itself (A_1, A_2, etc.) and waste products (Y), the cycle also produces membrane-forming molecules (T).

membrane, molecules T (Fig. 3.16). Hence the elementary steps of the cycle are

$$A_1 + X \rightleftharpoons A_2$$
$$A_2 \rightleftharpoons A_3 + Y$$
$$A_3 \rightleftharpoons A_4 + T$$
$$A_4 \rightleftharpoons 2A_1.$$

In general

$$\mathbf{A} + \mathbf{X} \xrightarrow{\;\;A\;\;} 2\mathbf{A} + \mathbf{T} + \mathbf{Y}. \tag{3.27}$$

Consider a spherical membrane consisting of n molecules T. Let us put the substances of our self-reproducing cycles inside it and assume that they cannot leave the space enclosed by the membrane but that the membrane is permeable to the source materials **X** and the waste products **Y**. Let the number of internal components of the cycle within the sphere be n. The incoming source materials X start the functioning of the self-reproducing cycle. During a single turn the cycle reproduces itself and also produces n molecules T:

$$n\mathbf{A} + n\mathbf{X} \xrightarrow{\;\;A\;\;} 2n\mathbf{A} + n\mathbf{T} + n\mathbf{Y}. \tag{3.28}$$

However, the n molecules T are incorporated into membrane $\boxed{T_n}$, thus reduplicating the surface of the membrane:

$$n\mathbf{T} + \boxed{T_n} \longrightarrow \boxed{T_{2n}}. \tag{3.29}$$

By combining eqns (3.28) and (3.29) we obtain

$$n\mathbf{A} + \boxed{T_n} + n\mathbf{X} \xrightarrow{\;\;A\;\;} 2n\mathbf{A} + \boxed{T_{2n}} + n\mathbf{Y}. \tag{3.30}$$

Thus the two systems are indeed connected to form a single supersystem; while one of its subsystems reduplicates itself, the other is also reduplicated.

But what kinds of processes do these equations represent? There is a spherule, something like a soap-bubble but of microscopic dimensions. It is not filled with air but with an aqueous solution and the whole spherule itself is immersed in an aqueous 'nutrient solution'. The components of the self-reproducing cycle are in solution within the spherule; they cannot pass through the membrane out of the spherule, but the components of the 'nutrient solution' can pass through the membrane into the interior space. Thus, as the nutrients enter the cycle begins to function. This process consumes the nutrients, but they are replaced from the external nutrient solution. Therefore a continuous nutrient current flows into the spherule.

But what are the nutrients used for? Some are used to form internal components of the cycle, which of course cannot escape from the spherule and so their quantity increases continuously. Some are used to synthesize of the basic substances of the membrane. These cannot escape from the spherule either, and are spontaneously incorporated into the membrane, i.e. the surface area of the membrane also increases. Hence both the internal contents and the membrane enclosing them are growing. Moreover, they grow at the same rate: reduplication of one occurs at the same time as reduplication of the other. Therefore while the content of the spherule doubles its surface area is also doubled.

Here we have a contradiction! If the volume of a sphere is doubled its surface area increases by a smaller factor, since the volume of a sphere increases in proportion to the radius cubed whereas surface area increases in proportion to the radius squared. However, the chemical 'constrained paths', i.e. the reactions, relentlessly produce just as many components of the internal cycle as membrane-forming molecules. There are three possible solutions to this problem. First, water flows from the external space to the interior of the spherule. Nothing prevents this, but the internal content becomes diluted and a difference in osmotic pressure arises, which acts to remove water from the internal space. Therefore this solution does not work. Second, the spherule becomes deformed, deviating more and more from the spherical and forming larger and larger relative surfaces. However, this is counteracted by surface tension and local electric charges. Therefore we are left with the third explanation: when the content and surface are doubled, the spherule divides into two spherules of equal size (Gánti 1974). There is nothing to prevent this, since the membrane is an elastic two-dimensional liquid, and the result is a doubled surface and a doubled volume.

But what does all this mean? We have combined two systems of a strictly chemical character into a 'super-system' (or, to put it another way, we have combined two chemical subsystems), and we have obtained a system with a surprising new property of expressly biological character. What can this system do? It is separate from the external world and its internal composition differs from that of the environment. It continuously consumes substances that it needs from the environment which are transformed in a regulated chemical manner into its own body constituents. This process leads to the growth of a spherule; as a result of this growth, at a critical size the spherule divides into two equal spherules, both of which continue the process. Thus after a given time the original spherule divides into two spherules, after double this time four spherules are obtained, after triple this time eight are obtained, then 16, 32, 64, 128, and so on, just as in the case of unicellular organisms multiplying by cell division (Fig. 3.17).

Would anyone observing such a spherule under a microscope, and realizing that it multiplies by division using nutrients, believe that it was not a living system? Can we believe that the spherule is not living?

G45The statement that there is no information-storing subsystem is based on the assumption that such a subsystem must include a template mechanism for the propagation of variation (mutations). If the mechanism is not template based, then the spherule described could be living according to Gánti's absolute life criteria. Suppose that certain kinds of molecules, which result from variant metabolic reactions and which can incorporate into membranes, are capable of 'recruiting' other similar molecules into small membrane plaques which chemically limit the movement of these molecules in the fluid plane of the membrane. Then the mutation consisting of the incorporation of one such molecule into the membrane will result in a hereditary change, namely a patch of variant molecules which can be transmitted through membrane reduplication. (Let us assume that there is an equilibrium amount of the molecule and that if membrane division results in no patch in one daughter spherule and an intact patch in the other, the patchless daughter will build a new patch after the metabolism has produced another variant molecule.) This kind of system is called a 'structural inheritance system' (Jablonka and Lamb 1995). The hypothetical example is similar to the cortical inheritance system in paramecia investigated by Sonneborn, who surgically removed pieces of cortex with cilia in one surface orientation and reimplanted them in different orientations. The modified cilia patterns were transmitted from parent to offspring for many generations through reproduction. Moreover, as a result of plaque formation, the membrane

(contd.)

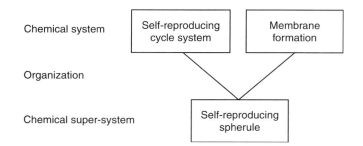

Fig. 3.17 A special chemical supersystem—a self-reproducing spherule—is formed by combining the functions of two distinct but compatible chemical systems, the self-reproducing cycle and the membrane-forming system. The supersystem thus formed displays qualitatively new properties, namely metabolism and the capability of spatial division.

A spherule, which is taking up nutrients, continues metabolism, is excitable, and multiplies itself! Moreover, since it always produces identical descendants, it seems to have the capability of heredity.

Nevertheless, the spherule is not a living system. It is not living, because it still lacks a fundamental property. This property is not one of the classical life criteria, but on the basis of knowledge gained from molecular biology it has been selected as an absolute life criterion. The spherule has no information-storing subsystem.G45 Hence this property is necessary for completion of the system. Thus we have to construct an information-carrying subsystem and integrate it into the system developed above. Then our system really will be living.

However, before doing this, we must express the results obtained so far in a formal manner, using equations. This is easy to do, since we only have to take into account the process of the division into two parts of the membrane surface $\boxed{T_n}$ doubled to $\boxed{T_{2n}}$:

$$\boxed{T_{2n}} \rightarrow 2\boxed{T_n}. \tag{3.31}$$

Thus eqn (3.30), generalized to nutrient materials and waste materials, becomes

$$n\mathbf{A} + \boxed{T_n} + n\mathbf{X} \xrightarrow[\text{①}]{A} 2n\mathbf{A} + \boxed{T_n} + n\mathbf{Y}. \tag{3.32}$$

However, this is not yet entirely correct, since $2n$A is also spatially separated into two parts. Each $\boxed{T_n}$ has its own nA and so they form a single system. Let this be denoted by brackets. Thus eqn (3.32) can be modified to

$$[n\mathbf{A}+\boxed{T_n}] + n\mathbf{X} \xrightarrow[\text{①}]{A} 2[n\mathbf{A} + \boxed{T_n}] + n\mathbf{Y} \tag{3.33}$$

which now expresses completely that $[n\mathbf{A}+\boxed{T_n}]$ is a supersystem composed of two parts such that its function, while consuming the nutrients nX, reproduces itself both quantitatively and structurally, i.e. the supersystem $[n\mathbf{A}+\boxed{T_n}]$ is multiplying.

3.10 The chemoton

The only thing missing from self-reproducing spherules, which prevents them from being living minimal systems, is an information subsystem. The two subsystems of the self-reproducing spherule correspond to two of the three cellular subsystems, i.e. the cytoplasm and cell membrane. Therefore if we can identify the third, informational subsystem and connect it functionally with the two, we may indeed be optimistic about arriving at the minimal system of life.

Again, we must start by defining a well-known and frequently used concept—the term 'information'. In what follows, we shall not be using this word in its usual very wide sense, particularly when we are discussing information-carrying systems.

Information is carried by all existing objects. A stone on the roadside carries information about its matter, its chemical and mineral composition, its circumstances of generation, etc. The same statement is true for any object. However, human culture produces systems which carry, as well as information about themselves, another type of information to which they are not so intimately and inherently connected. For instance, this book carries not only information about itself (e.g. that it is made from paper, it consists of pages, and has a cover), but also carries (or at least the author hopes that it carries) information about the principles of living systems, which is nothing to do with the book as an object.[G46]

Books and other written texts are not the only objects carrying surplus information of this kind. Magnetic tapes, punched cards, electronic storage devices, etc. are also capable of storing information with which has nothing to do with them as objects. When considering the storage of information in the following discussion, we shall always be concerned with this 'surplus information', and information-storage systems are understood to be capable of retaining it.

A common property of information-storage systems of this kind is that information is stored using signs. Anything can function as a sign: a point, a dash, a letter, a hole, a knot, a state of magnetization, etc. Almost anything which is available in sufficient abundance and which can be fixed in space and in time so that it differs from its environment in order that the information it represents can be deciphered by an appropriate information-reading system can be used as a sign. Information becomes effective information if, and only if, another system is capable of reading and using it.

Information can be stored using any kind of sign, namely by the quantity and quality of the signs as well as by the order of different signs.[G47] In the case of the first three figures of the Roman numerals the value of the number, i.e. the information to be communicated, is expressed by the quantity of signs, in the case of the first 10 ciphers of the Arabic numerals it is expressed by the quality of signs. In both numeral systems information concerning the value of higher numbers is coded by the order of signs of different qualities (Fig. 3.18).

contains at least a small amount of 'surplus information' in Gánti's sense developed in the next chapter, e.g. information about the state of metabolism since membrane size carries information about the number of metabolic cycles of the specific reactions which produced the molecular constituents of the membrane. Thus the variant membrane is also an information-storing system. This would be a system of 'limited heredity', in Szathmáry's sense, rather than an unlimited one like the template-sequence system of modern nucleic acids.

Nevertheless, in principle, a hereditary substance in the form of a templating molecular polymer is not necessary for an information-storing subsystem. It has been speculated that the branched carbohydrate molecules incorporated into the membranes of modern cells might offer sufficient complexity to constitute a (nearly) unlimited hereditary system (Sharon and Lis 1993).

[G46]Philosophers call this 'surplus' information 'semantic content' or 'content' for short. It is what the symbols are 'about' rather than about the symbols themselves. See Dretske (1981) on the role of information in semantics and epistemology.

[G47]The general theory of signs, which is widely considered the foundation of semiotics, was developed and articulated by the polymath and philosopher Charles Saunders Peirce (1897).

I II III   MDCXXXII

0 1 2 3 4 5 6 7 8 9   19528295

Fig. 3.18 The first three Roman numerals express the information to be communicated (the numerical value) by the simple quantity of the sign, the first 10 Arabic numerals express it by the quality of the signs. In both systems information concerning higher numerical values is coded by the sequence of different signs.

It is easy to see that chemical information can also be stored by the quantity and quality of signs. The hormonal regulation of multicellular organisms and the informational role of certain aromatic substances (the pheromones) in ant societies depends mainly on the quality of signs. However, alluring aromatic substances emitted by the female butterfly do not inform the male by quality alone, but also by quantity; the male is attracted to greater concentrations of the aromatic substances and thus can find the female even at a distance of several kilometres.

Chemical communication systems like these can only transfer instructions of a directing character necessary for regulation, and thus they are capable of storing information for only a short period. Moreover, this type of storage is unsuitable for more complicated information because the signs, i.e. the individual molecules, are in constant unconstrained motion. Thus they cannot be used to form an order of signs which would allow a large information-storage capacity. Therefore if complex information is to be stored chemically for a long time, the signalling molecules must be fixed in space so that the information stored in them is available for soft (chemical) decoding systems.

Molecules as signs can be fixed by chemical bonds, i.e. by polymerization. If we denote molecules serving as signs by V, then chemical information fixing via the quantity of the signs can be expressed by the polymerization equation

$$n\text{V} \rightarrow p\text{V}_n \tag{3.34}$$

where n is the number of molecules used to store the information (or, expressed chemically, the 'degree of polymerization') and $p\text{V}_n$ is the polymer molecule formed from n molecules V.

The polymer may have either a stereo-network or a linear structure. Polymers of the former structure are obviously inappropriate for chemical information storage because the individual signs are not available for chemical decoding mechanisms as they are more closely packed. In contrast, linear polymers are ideal for information storage, since the signalling molecules, which are ordered like beads on a string, can be accessed from any direction except the line itself.

However, even the polymerization process given by eqn (3.34) leads to a linear macromolecule, this property alone is not sufficient for effective information storage. In this case information would be stored by the simplest means, i.e. by the quantity of signs. However, what determines the length of a linear polymer, i.e. the number of signs built into the individual polymer macromolecules? In a common polymerization reaction mixture the polymers formed are not of equal lengths—some are very short and others are very long. However, information storage can only be achieved by linear polymers in which the number of the molecular signs is exactly determined. In this simplest case just the number of the molecules carries information within the macromolecule.

However, there is a special type of polymerization—template polymerization—in which not just any kind of macromolecule is formed. The word 'template' really means 'pattern', as this process has a type of 'patterned' structure. In this process polymerization only occurs if a polymer molecule is already present—a 'polymer file' serving as the template and binding onto its surface the individual sign molecules which then become bound to each other. Thus the type of macromolecule that will be formed is determined by the template. However, the new macromolecule can also serve as a template. Hence the number of templates is doubled and there are now twice as many template molecules present, i.e. twice as many macromolecules can be synthesized at the same time. Since they can also serve as templates, the number of polymer molecules present in the reaction mixture will be four times, eight times, 16 times, etc. the number of templates originally present.

Again, we have an autocatalytic process—the template molecule is an autocatalyst! This fact in itself suggests that the process of template polymerization can easily be connected to the other two autocatalytic chemical systems discussed in previous sections. However, the template mechanism is not just useful with respect to information storage. The mechanism for decoding is also included, since the information present in the template must be read when it is used for the synthesis of a new macromolecule.[G48] Moreover, not only do we have a reading mechanism, but also a method for multiplying the information because the new molecule contains the same information as the template from which it was formed. Thus template polymerization appears to be appropriate for the role of the third subsystem in the minimal system of life.

Perhaps we should remind the reader that our aim is to find the simplest possible theoretical system, regardless of whether it can be actually realized. However, as with the other two subsystems, we shall also show here that this system is not just a product of fantasy; in fact, template polymerization systems can be found in the real world. Primarily, the well-known process for synthesizing DNA is such a system, as is RNA synthesis as mentioned previously. In living organisms these template polymerization processes occur by means of enzymes, but there is experimental evidence to show that the template effect may also take place in non-enzymatic nucleic acid synthesis. The Hungarian researcher Ferenc Cser and his colleagues found that polymerization reactions of this kind also occur in non-biological compounds, specifically in the polymerization of acenaphthilene. Thus, taking into account the role of the pattern (the template), we can now rewrite eqn (3.34) as

$$n\mathrm{V} + p\mathrm{V}_n \rightarrow 2p\mathrm{V}_n \qquad (3.35)$$

[G48]Since the information is only coded by the length, not the sequence, of the polymer, 'reading' amounts only to the syntactic detection of monomers until the total length is reached. Like reading a long book consisting of a single word, the semantic information received is very limited.

which includes the autocatalytic character of the process in addition to its information-storage properties. Thus we have everything required for the construction of the abstract model of the theoretical simplest living system—the chemoton. Moreover, we have a ready recipe for it in the form of the self-reproducing spherule. It is now necessary to establish a 'supply and demand' connection between the different chemical systems, i.e. to select a self-reproducing cycle which can produce the raw materials necessary for the functioning of the informational system.

As the simplest solution, let our self-reproducing cycle consist of a sequence of chemical reactions, each of which performs a single task. Thus the first step makes use of the nutrients:

$$A_1 + X \rightleftharpoons A_2. \tag{3.36}$$

Waste materials are produced in the second step:

$$A_2 \rightleftharpoons A_3 + Y. \tag{3.37}$$

The third step is the production of the information-carrying molecules:

$$A_3 \rightleftharpoons A_4 + V. \tag{3.38}$$

The fourth step is the formation of the membrane-forming molecules:

$$A_4 \rightleftharpoons A_5 + T. \tag{3.39}$$

Finally, the fifth step is self-reproduction:

$$A_5 \rightleftharpoons 2A_1. \tag{3.40}$$

Now we have a self-reproducing cycle consisting of five steps which also produces the basic materials for the other two subsystems (Fig. 3.19). The equation for this cycle is obtained by combining eqns (3.36)–(3.40):

$$A_1 + X \xrightarrow{\quad A_1 \quad} 2A_1 + V + T + Y \tag{3.41}$$

or, more generally,

$$\mathbf{A} + X \xrightarrow{\quad A_1 \quad} 2\mathbf{A} + V + T + \mathbf{Y}. \tag{3.42}$$

If this system operates in the interior of a membrane spherule containing the information-carrying macromolecule, the sign molecules V produced here are polymerized according to the template present, and

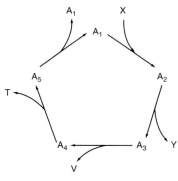

Fig. 3.19 Self-reproducing minimal cycle for the chemoton. In addition to its own constituents (A_1, A_2, etc.) and waste materials (Y), the cycle also produces the raw materials for the informational subsystem (V) and membrane formation (T).

the membrane-forming molecules T are built into the membrane itself. Thus the equation for the chemoton will be obtained if eqn (3.33), which describes the function and multiplication of the membrane spherule, is connected with eqn (3.35), which characterizes the working of the chemical information system (i.e. by combining the two equations):

$$\left[n\mathbf{A} + p\mathbf{V}_n + \boxed{\mathbf{T}_n} \right] + n\mathbf{X} \xrightarrow{\overset{A}{\textcircled{1}}} 2\left[n\mathbf{A} + p\mathbf{V}_n + \boxed{\mathbf{T}_n} \right] + n\mathbf{Y} \qquad (3.43)$$

where the expression

$$\left[n\mathbf{A} + p\mathbf{V}_n + \boxed{\mathbf{T}_n} \right]$$

denotes the chemoton which, in the interior of the membrane spherule built up from n molecules T, contains the minimal system corresponding to the protoplasm, i.e. the self-reproducing cycle nA, and the information-carrying polymer $p\mathbf{V}_n$ corresponding to the genetic material.

Obviously, eqn (3.43) implies that the three chemical systems work in complete harmony: they grow at the same rate, they reproduce simultaneously, and, as in the case of the self-reproducing chemical spherule, after reaching a critical size the three systems are divided into uniform, functioning, and self-divisible systems, i.e. they multiply. The three chemical systems form a single supersystem, in which the function of the original chemical systems is subordinate to the function of the integral supersystem.

Now, we have finally obtained a minimal system which corresponds to the basic organization of the prokaryotic cell: it contains the subsystems corresponding to the roles of the cytoplasm, the genetic material, and cell membrane, and it is capable of functioning, growing, and multiplying. From a chemical point of view, we have to consider this as a supersystem, since we have constructed it from different chemical systems. However, from a biological point of view it is clearly a minimal system, since it is the theoretically simplest system which displays all the principal life phenomena. Thus the chemoton can be considered to be the basic unit of life. However, we still have to prove that the chemoton is effectively living.

3.11   Life at the chemoton level

It will not be difficult to prove the living character of the chemoton, since we have already discussed in detail the life criteria which must be fulfilled in order that a system can be considered to be living. We now have to determine whether the chemoton satisfies these criteria. If it satisfies the 'actual' life criteria, we can consider it to be living. However, it would be even better if we found that it also satisfied the 'potential' life criteria. In this case we would hope that it could be considered the basic unit of an exact theoretical biology, from which the abstract theoretical model system for the whole living world could be deduced. Of course, the exact proof cannot be presented in this book, but the qualitative validity of our statements can easily be clarified by the following simple considerations.

First, let us consider the life criteria. According to the first life criterion the system must be an inherent unit which cannot be constructed by adding together the properties of its parts, or subdivided into parts which carry properties belonging to the system as a whole. It is unnecessary to prove this criterion separately since in the construction of the chemoton parts (subsystems) the connections resulted in new qualitative properties which were not present in the individual subsystems. Conversely, the chemoton cannot be separated into parts which display the properties of the whole. The special properties of the chemoton are due exactly to the compositional mood of the three subsystems. Hence the chemoton satisfies the first real life criterion.

The second criterion requires that a living system should perform some kind of metabolic process. By this we understand that matter and energy enter the system, either actively or passively, from the external environment and are then transformed into the system's own internal materials with the production of waste. These chemical reactions lead to a regulated and controlled growth of the internal substances as well as to supplying energy to the system. Finally the waste materials leave the system, either actively or passively.

The chemoton can perform all these processes. An energy-rich nutrient X moves into the system from the environment and the chemical reactions of the system transform it into A, T, and V, which form membrane and genetic material in a regulated (and, as will be seen, controlled) manner. Waste material Y is also obtained and can leave the system through the membrane which is permeable to it. Moreover, the 'chemical motor' of the system is driven by the energy released during the transformation of the nutrient X, and the functioning of the system is ensured by this released energy. Therefore the chemoton satisfies the second real life criterion.

According to the third criterion, the living system must be inherently stable. This inherent stability is ensured by the inner organizational mode of the 'chemical motor' subsystem, which operates cyclically. Hence this property is inherent in one of our subsystems. Thus the chemoton satisfies the third life criterion. Moreover, chemotons are remarkably adaptable to environmental changes and are capable of compensating for them quite effectively, as will be shown by computer simulations in the next section.

The fourth real life criterion states that living systems must contain a subsystem which carries information concerning the total system. We already know that the chemoton has a subsystem capable of information storage, but we still have to prove that this information refers to the integral system of the chemoton.

The information-storage role of the chemoton can easily be established for its more complicated variants, the mutative chemotons, which will be dealt with in a subsequent section. In the next section, we shall describe a computer simulation experiment which demonstrates that this characteristic can also be found in the simplest form of chemoton. However, we shall outline the basic principle here.

The chemical reactions of the self-reproducing cycles considered so far are reversible, i.e. they can be driven forwards or backwards depending on the given conditions. Thus such a cycle can operate in both directions. If there is an excess of nutrients, it will go 'forward' and thus will become self-reproducing. However, if there is a lack of nutrients and excess waste material in the mixture, the cycle may operate in the opposite direction and become self-consuming. Which of these happens and the rate at which it happens depends primarily on the concentrations of the nutrients and the waste materials, of course, the cycle may also be stationary as a function of the concentrations.

In living systems, however, the 'chemical motor', i.e. the complete reaction sequence, cannot operate in the reverse direction—cell growth is irreversible. Of course, the cell may decompose but this can never occur along the same path as its growth. Thus irreversibility must be ensured in the complete chemoton system, and this can be done by incorporating irreversible processes.

This does not present many difficulties. The self-reproducing spherule itself contains such a unidirectional process, namely membrane formation. The membrane-forming molecules are spontaneously built into the membrane, but are not able to break away spontaneously. Thus in the self-reproducing spherule the cyclic system can only work in one direction, namely towards constant self-reproduction. This is because one of its products, the membrane-forming molecule T, can never be accumulated as it is immediately and spontaneously incorporated into the membrane. Thus the self-reproducing spherule can only work in the direction of growth and multiplication.

Template polymerization is also a unidirectional process, and is appropriate for ensuring the irreversible functioning of the system. Unlike the membrane-forming molecules T which are immediately incorporated in the membrane, the sign molecules V only bind to each other according to the pattern of the template molecule when they have reached a critical concentration. Thus the accumulated V molecules may slow down the reaction, or even stop it. (Obviously, they cannot reverse the reaction since this would also require the presence of T molecules which have been incorporated in the membrane.)

When the concentration of V molecules reaches the critical value necessary to start template polymerization, most of them will be consumed in the formation of the new polymer molecule pV_n. Thus the concentration of V decreases and the cycle functions more rapidly. The decrease in the concentration of V is a function of the length of the template molecule, i.e. of the number n of sign molecules present in the template pV_n. A longer template molecule (i.e. a larger n) requires more molecules V for its reproduction, thus decreasing the concentration of V and allowing the cycle to function more intensively. Conversely, a shorter template molecule allows the cycle to function more slowly and less intensively. As will be shown by the computer simulations, this has a profound effect on the properties of the chemoton. Furthermore, these properties are hereditary, since the reproduction of the chemoton involves the reproduction of pV_n in its characteristic and constant length. Thus the template molecule pV_n carries information concerning the system as a whole, which means that the chemoton also fulfils the fourth real life criterion.

According to the fifth life criterion, processes in the living system must be regulated and controlled. The previous discussions concerning the fourth criterion and metabolism form a sound basis for assuming that the various substances in the system are produced according to definite relations and in a regulated manner. In view of the way that the integral system functions, either pV_n or the membrane may have a controlling role. Thus the chemoton may fulfil the fifth life criterion.

The sixth life criterion (the first potential life criterion)—growth and multiplication—need not be dealt with separately, since we have already given a detailed demonstration that this is a property of a self-reproducing spherule. Hence the chemoton also fulfils the first potential life criterion.

The seventh life criterion (i.e. the second potential life criterion) is the capability for hereditary change. When we deduce some of the fundamental laws governing heredity in the section on mutative chemotons, it will be seen that the chemoton has this property. Nevertheless, the fundamental reasons for this can already be seen here. It was mentioned above that the hereditary properties of the simple chemoton are determined by the length of the macromolecule pV_n. Template polymerization is subjected to errors: shorter copies may be formed or two pV_n molecules may be connected to a molecule of double length, resulting in molecules with a different length serving as templates. This will produce descendants with the changed property, which will be inherited from generation to generation. Thus chemotons satisfy the second potential life criterion.

The eighth life criterion (the third potential life criterion) is the capacity for evolution. Computer simulations prove that the chemoton satisfies this criterion.[G49]

The final criterion was the criterion of death, although not a 'programmed' death as unicellular organisms are potentially immortal. Cellular death means that there are natural circumstances in which the structure of the cell deteriorates and the cell perishes. Chemotons have a kind of mortality just like this. Their membrane system is extremely sensitive, and if the membrane surface deteriorates the chemoton perishes. Thus chemotons also satisfy the final life criterion.

Thus we can state that the chemoton is indeed an abstract minimal system carrying all the fundamental and characteristic general qualitative properties of living systems. Moreover, it fulfils the potential life criteria, and thus we can assume that the chemoton is an abstract fundamental unit of life from which an abstract model system concerning life,

[G49]Demonstration that something is a unit of evolution does not require computer simulation, nor can such simulations be sufficient because the demonstration is a logical problem. Whether units of evolution are taken to be entities exhibiting heritable variance in fitness (Darwin's principles), or multiplication, variation and, heredity (Maynard Smith 1987), or something else, the problem is to determine whether the entities in question have the required properties. Therefore, unless the properties concern dynamic behavior directly, computer simulations cannot show that the required properties are present. At most they can show that the system behaves *as if* it has them. Because Gánti has already shown that chemotons multiply, can vary, and have heredity, they satisfy Maynard Smith's criteria for units of evolution. Showing further that their variations are correlated with fitness would be sufficient to show that they are units of selection in the sense of Lewontin (1970) and Darwin.

i.e. an exact theoretical biology, can be deduced with mathematical precision.

3.12   Computer simulation of the function of chemotons

Our equations so far have been stoichiometric, i.e. they described what kinds and quantities of substances could be found in the given solution before and after the occurrence of the events denoted by the equation. However, these equations do not contain any information about the actual occurrence of the events or about the rate at which they occur. They cannot tell us whether seconds or millions of years are necessary to complete a process, or anything about the behaviour of the system under the influence of external circumstances. Despite this, we have dared to deduce from them all kinds of expressly kinetic conclusions about metabolism, regulation, control, growth, multiplication, etc. Did we have the right to do this and, if we had, how can these statements be proved exactly?

The description of kinetic events, i.e. of the behaviour of dynamic systems, requires differential equations describing the course of the events as a function of time. Differential equations describing the progress of a reaction, i.e. its 'rate', were formulated a century ago in chemical, or reaction, kinetics. Hence it is not difficult to write down the appropriate differential equations for the individual reactions of the chemoton. However, two problems remain.

First, a differential equation itself does not provide information about the behaviour of the system; all it provides is a prescription for the mathematical solution of the equation, i.e. for its integration. The self-reproductive steps within the individual reactions of the chemoton are described by second-order non-linear differential equations where the general mathematical solution is unknown. Thus the complete system of equations for the chemoton is insoluble in the closed form and provides no information about the working of the chemoton.

However, the system of differential equations can be solved by the method of approximation which, because of the very lengthy computations necessary, can only be performed using a computer. Therefore, with the help of Ferenc Békés and Ákos Nagy, we have performed many computer investigations 'simulating' the functioning of the chemoton. We have obtained very interesting results, which will be discussed below. However, before doing this, we must consider the second problem mentioned above.

A numerical approximation can only be performed for given numerical values. In our case, this means that we cannot obtain a general description of the functioning and behaviour of the chemoton. Only the behaviour of a specific chemoton, which has a specific composition and function for given specific circumstances. Obviously, innumerable internal compositions and an infinite variety of external circumstances can be envisaged, and in order to understand fully the behaviour of the chemoton we would have to perform an infinite number of computer simulations.

Of course, it is not necessary to calculate every possible case—a reasonable choice of conditions allows the number of calculations to be reduced to a reasonable level. However, even with these restrictions the computational capacity required would be much greater than that actually available.

So what can be done? Some computer simulations of the capabilities of growth, multiplication, evolution, etc. must be performed, while the general consequences are, as far as possible, determined from the general stoichiometric equations discussed earlier. This can be done because the stoichiometric equations are not just simple expressions of a material equilibrium. They can be used to formulate certain kinetic results, if a slight 'trick' hidden in them is used.

This 'trick' is the sign of the cyclic process. In 'honest' stoichiometric equations the two sides are related by the sign of equality, often substituted by a double arrow. Stoichiometric equations describe what kind of substances are formed (quantitative relations) from what other substances if the process is completed. The signs of cycles introduced by us denote this completion, but also imply that the cycle may be repeated with the same results.

Thus our cycle signs do not inform us about events within a period, since they give only the final result of the period. However, if the duration of the period is chosen as the unit of time, they show what happens during an arbitrary sequence of such time units. With respect to our chemoton this means that the equations do not inform us about the behaviour of the individual chemoton, but they describe what kind of chemotons are produced by a chemoton, i.e. what happens to the chemotons during the sequence of generation. Given the duration of the periods (the 'generation time'), it is possible to determine what happens from the function of common physical time. Thus a stoichiometric equation with a cycle sign can be treated as a kinetic equation with a single period as the lower limit of its 'resolving power'.

Nevertheless, it must be remembered that stoichiometric equations always give the result of a reaction that has been completed, i.e. they refer to a complete irreversible state. However, most (in principle all) of the chemical reactions are reversible processes tending to an equilibrium which they never completely reach. This is well known to chemists, but nevertheless they write down stoichiometric equations for these reactions, since these are very easy to deal with in practical computations and, if the equilibrium state is known, the results can easily be corrected.

Strictly, the stoichiometric equations given for chemotons are valid only if each of them is irreversible. However, we have required that most of the reactions are reversible, leading to an equilibrium. In the case of highly complex systems like this, it is not as easy to correct for deviations as in a simple reaction. Hence the actual composition of the chemoton depends on the conditions existing and can only be determined by computer simulation.

At the beginning of our investigations, when our initial primitive simulations did not take into account the proper functioning of this highly complex chemical system, we expected the computer to give the same results for the composition of the chemoton as were obtained with the stoichiometric equations. However, we obtained quite unexpected and apparently incomprehensible results. As always in similar cases, we looked for the reason for the failure, and as we did not find it, we began to suspect our mathematician colleague of erroneous programming. Of course, he protested and a thorough investigation proved that the program was faultless. Our chemoton could do more than we expected— more than we had dared even to hope. (And this was not the only surprise caused by the chemoton!)

We found that, in the chemoton (as in living organisms), inherited properties are not expressed rigidly but rather are modified in accordance with the effects of the environment. Even when endowed with the same hereditary properties—the 'genotype'—an individual can be fat or thin, strong or weak, etc., depending on the nutrients available and the effects of the environment. Only when we realized this did we understand properly that heredity describes the identity of properties for subsequent generations only under identical conditions (in the living world and in chemotons alike). Within the still very wide possibilities determined by inheritance, different environments may produce rather different individuals.

The adaptability of the chemoton results from the reversibility of the chemical reactions—from their attempts to achieve an equilibrium. If the chemoton consisted only of irreversible steps, heredity would be expressed without adaptability and the results could be read exactly in the stoichiometric equations. However, reversibility generates adaptability, as we shall see from the simulation experiments.

Concerning simulations, one often encounters the opinion that 'a computer can only give back what was put into it'. In a certain sense, this is of course true. However, we did not feed anything into the computer about the functioning of the chemoton! So how did it produce the above-mentioned results? This 'serendipity' was just the essence of our computer simulations.

As already mentioned, it has been known for a long time that the rate of a reaction can be described by a differential equation. For instance, let us consider the first chemical reaction in the chemoton cycle:

$$A_1 + X \rightleftharpoons A_2.$$

If we use a computer program which integrates the differential equation corresponding to this equation, we obtain the course of the reaction, i.e. the quantity of the reaction product A_2 in the system at any given moment. Obviously, this value depends on the external conditions of the reaction (e.g. temperature) and on the concentrations of the components (e.g. X), among many other variables. Now we can obtain from the computer the results that we wanted. For example, if we add a large quantity of X, then a large quantity of A_2 will be formed at high rates, but if we feed in a small quantity of X, then a small quantity of A_2 will be formed at a slow rate. But the same thing happens in a test-tube if we add a large or a small quantity of component X. Hence the computer can tell us how the chemical reaction would take place within the conditions assumed in the test-tube itself.

Of course, the same argument holds for the second reaction of the cycle:

$$A_2 \rightleftharpoons A_3 + Y.$$

Here too the course of the reaction can be arbitrarily chosen, within the constraints set by the equation.

However, if the differential equations corresponding to the two reactions are fed into the computer together, so that the reaction product of the first is the raw material of the second, i.e.

$$A_1 + X \rightleftharpoons A_2 \rightleftharpoons A_3 + Y,$$

the second reaction can no longer run in an arbitrary manner because the quantity of raw material produced by the first reaction will determine its course. The course of the first reaction will also be restricted, because it depends on how much of the end-product is removed by the second. This is true for computer calculations of both simulated processes and chemical reactions. Two reactions 'run down' in a different manner if they are connected than if they are not. Since both influence and control each other, they depend on each other.

In computer simulations of the chemoton, only the differential equations of the individual chemical reactions were put into the computer; no instructions were given concerning the functioning of the chemoton as a whole. We let the computer calculate the process which would have occurred in a solution where several connected chemical reactions ran simultaneously, and we registered how these reactions influenced each other's courses.

First it was confirmed that, as expected, the system is capable of functioning—the chemoton can maintain metabolism, i.e. take up nutrients. It can make the whole system work from the energy of the nutrients and transform them in a regulated and controlled manner into the reactants required by the system so that it can grow continuously.

Next, it was found that the system is excitable and sensitive, i.e. it reacts sensitively to external changes (e.g. changes in temperature, changes in nutrient concentration). This reaction manifests itself by a change in the internal composition of the system—the proportions of its internal components are changed. However, the system also reacts sensitively, and during its individual life, i.e. from division to division, its composition changes in a prescribed manner. Given identical external conditions, the same changes are then repeated in each of the offspring during their individual lives.

Then it was proved that the information storage system (the macromolecule pV_n) indeed has a controlling and inheritance-determining role in the function of the chemoton. The length of pV_n, i.e. the value of n, determines what types of internal changes occur and how they occur during the individual life of the chemoton. Since the length of pV_n in the descendants is identical, the same changes will occur in every offspring during its individual life.

We discovered with surprise during our computer simulations that, within the chemoton, pV_n plays a role similar to a pacemaker. The reduplication of the pV_n occurs only once between two divisions; in the individual life of a chemoton the synthesis of pV_n occurs only once, when it is prepared to divide according to the external circumstances. The occurrence of pV_n synthesis produces a physical sign through the decrease in osmotic pressure to begin the process of division. Thus pV_n not only controls the rate of metabolism, but can also control the onset of the division, even in this simplest model serving as a minimal system.

However, our greatest surprise came from the simulation investigations of the stability of chemotons. The chemoton is after all a chemical system, since every process in it, except division, is a chemical reaction. Now, the rate of the chemical reactions depends on the concentrations of the raw materials; to a rough approximation it is proportional to the concentrations of the raw materials. Therefore we performed simulation experiments in which the concentration of the nutrients was suddenly decreased to a tenth of its original value. According to well-known chemical experiments we expected the metabolic processes of the chemoton to take place 10 times more slowly, i.e. for the generation time to increase 10-fold. For instance, if a chemoton need 1 hour to reduplicate itself under the original conditions, we expected that, after reducing the concentration of the nutrients to a tenth of the original, the reduplication time would be increased to 10 hours. However, to our great surprise, the generation time increased by only 10 per cent. Hence a tenfold decrease in the nutrients influenced the functioning of the chemoton 100 times less than expected on a purely chemical basis. More detailed investigations revealed that, after these violent environmental changes, the internal composition of the chemoton undergoes substantial rearrangement; the quantity of the internal component reacting directly with the nutrients (A_1) is increased by a factor of 10 at the cost of the other components, thus compensating for the decrease in the external nutrient. Hence the reaction rate remains substantially unchanged.

Thus chemotons have a considerable ability to compensate for external change; they can make their internal functioning largely independent of external circumstances.

Hence our computer simulation experiments supported our previous conclusions drawn from the stoichiometric equations for the chemoton. They proved that chemotons are individual, functioning, metabolizing, inherently stable, homeostatic, excitable, regulated, and controlled systems which can grow and multiply, are provided with information which can be inherited, and are capable of hereditary change.

Our conjecture that the chemoton is an abstract living unit which can appropriately be taken as a basis for a truly exact theoretical biology has now been strengthened. It is worth citing Pavlov again:

Life as a whole, from the simplest organisms to the most complicated ones—man of course included—is a long sequence of counterbalancings with the environment, a sequence which becomes eventually very complex. The time will come—even if it is still rather far away—when mathematical analysis supported by scientific analysis can put into majestic formulas of equations these counter-balancings putting finally into an equation life itself.

Is it be too daring to state that the time has come for the fulfilment of this task? Let us try anyhow, on the basis of our previous considerations.

3.13   Chemoton theory involves the principles of soft automata

Descartes considered animal organisms to be automata. His mechanistic ideas were later criticized, mostly legitimately. Nevertheless, his ideas were based upon real observations: he perceived that machines and living organisms have common properties. Obviously, these common properties could not be expressed exactly in his time, since neither the functioning of living organisms nor that of machines was understood. Therefore a search for the common properties of (and basic difference between) machines and organisms is not performed by following Descartes' ideas but on the basis of modern scientific knowledge.

Machines (and other devices of modern technology, the 'automata') and living organisms are both organized systems whose characteristic qualitative properties originate from the manner of their internal organization. Automata and living organisms are both dynamic systems; their characteristic properties present themselves during function. Thus they need an energy source, and both can transform external energy into directed useful work. The transformations within them are not arbitrary, but take place in a regulated and controlled manner.

It was not by chance that Descartes considered living organisms to be related in some way to machines, since natural systems with properties characteristic of machines can only be found in the living world. Only in living systems can these properties be found together. Moreover, all animals have the property of active 'self-controlled' movement, which shows an extraordinary similarity to machine movement.

Nevertheless, the differences between organisms and machines are still more important than the similarities. First, living beings are soft systems, in contrast with the artificial hard dynamic systems. Furthermore, machines must always be constructed and manufactured, while living beings they construct and prepare themselves. Living beings are growing systems, in contrast with technical devices which never grow after their completion; rather, they wear away. Living beings are multiplying systems and automata (at least at present) are not capable of multiplication. Finally, evolution—the adaptive improvement of living organisms—is a spontaneous process occurring of its own accord through innumerable generations, whereas machines, which in some sense may also go through a process of evolution, can only evolve with the aid of active human contribution.

These are not negligible differences. In fact, the differences between living beings and automata are qualitatively more fundamental than the similarities. Nevertheless, this does not mean that the similarities are not based upon common principles; identical phenomena in machines and in living systems have identical laws in their background, i.e. the laws of cybernetics and automata theory. If this is the case, then elements and basic laws of cybernetics can be applied not only to the nervous system, neural networks, neural regulation, etc. (which at their own levels can be considered hard systems, as they are organized by fixed spatial connections), but also to soft chemical systems and processes occurring in solution. This is exactly what is done in chemoton theory, and so in attempting to disclose the basic principles governing the function of soft (chemical) automata[S32] we have only to consider the results discussed earlier in this section from the point of view of the theory of regulation and control.

However, this is not an easy task. Machines and automata are constructed from well-known dynamic elements such as sensors, switches, relays, valves, wheels, wires, modulators, amplifiers, difference-formers, sign-formers, etc. All of these can be realized practically in many different ways. For instance, sensors include tripping circuits, voltage regulators, tachometer dynamos, governors, tripping transformers, string galvanometers, rheostats, thermocouples, contact thermometers, Bourdon tubes, liquid manometers, pressure gauges, Pitot tubes, bolometers, flowmeters, etc. Difference-forming devices include instruments such as amplifiers, differential manometers, differential gears, feeder voltage regulation transformers, Wheatstone bridges, etc. It is easy to see that machines, devices, and automata can be constructed from such elements, provided that they are appropriately chosen and connected. But how can all this take place in a solution? How can switches, amplifiers, wires, etc. exist in a beaker of water and where can these be localized as organs? Obviously, the stimulus must somehow be conducted from the sensor organ to the effector organ, but how is this possible if the sensor, conductor, and effector are all dissolved in a beaker of water?

Nevertheless, it is possible. We cannot present here the details of the functioning of soft (chemical) automata, but we can give the basic principles by which such automata function in the living world. Let us now consider them one by one.

First, we need some kind of a sensor. Sensors can be anything with properties which are modified by environmental changes, provided that this change of property can influence the function of the machine or automaton. Obviously, the electrical properties of the sensor must be capable of change in an electric device, while in a mechanical device the sensor must go through a property change of a mechanical nature. Thus

[S32]A complementary approach was provided by Rössler (1972) in a German paper entitled: 'Basic switches of fluid automata and relaxation systems'. The crucial difference between his approach and that of Gánti is that enzymes play no role in the latter (at least in the basic formulation). See also Rössler (1974).

the sensor of a chemical automaton must undergo a change in its chemical properties. Furthermore, this change must depend on the magnitude of the external effect and must be reversible, i.e. as the external effect decreases or disappears, the chemical property should also change back or return to the original position. In chemistry, this requirement is ideally satisfied by chemical reactions leading to an equilibrium, i.e. the reversible reactions.

When equilibrium has been reached in a reversible reaction

$$A \rightleftharpoons B$$

the ratio of the number of molecules of A to the number of molecules of B is constant under the given conditions, and this constant (the 'equilibrium constant' K) is characteristic of the reaction. In this definition we must emphasize the 'given conditions'; if the conditions change, the numerical ratio of A to B also changes, and if the original conditions are then restored, the original ratio will also be restored. Of course, the word 'condition' can mean many things: temperature, pressure, concentration, pH, or anything registered as a 'state variable' in the thermodynamics of aqueous solutions.

Thus we have here an excellent sensor, from more than one aspect. Firstly, it is capable of measuring not just one but many state variables. It is reversible and works without friction, and therefore it does not break down. Neither can it err, because the end result is the statistical sum of the transformations of the individual molecules and the number of the molecules is so enormous (generally 10^{20}–10^{22} per litre) that the possibility of error is practically excluded. Compared with the elements of hard automata, these are advantages which would open up incredible possibilities for technological applications, and it is just these possibilities that have been exploited by the living world. The soft automata of living systems are almost error free; the 'hard' automata in living organisms are always responsible for their functional mistakes.

However, we have run a little ahead of ourselves. So far, we have only discussed the advantageous properties of a single constituent, the sensor, but we have already extended this to the advantages of soft automata in general. Now we have to prove that the previous considerations also hold for the other constituents and, in general, for the whole of soft automata.

A reversible reaction perceives environmental changes in the sense that under their influence its equilibrium will be changed. The quantity (concentration) of component A increases at the cost of component B, and vice versa, like the pointer of a balance or a measuring instrument when it is deflected from its original position. The new equilibrium corresponds to the new conditions, but how does the whole of the system register this change?

In the case of a technical automaton the effect passes through a sequence of component parts to the effector, i.e. through the action chain.

Chemical systems may also contain such action chains, namely the reaction chains. Indeed, the product of one reaction serves as the raw material for the next, and the product of this reaction will be the raw material for the following one, and so on. Thus we have an action chain of chemical reactions which convey the action through elementary reaction steps from the sensory chemical reaction to the required chemical transformation. Thus in a reaction chain

$$A \rightleftharpoons B \rightleftharpoons C \rightleftharpoons D \rightleftharpoons E \cdots,$$

where the product of one reaction is the raw material of the next, a disturbance—or a 'sign'—produced anywhere in the chain will run through the whole of the chain. Moreover, this conduction of a disturbance, action, or sign does not occur between two determined points in space, as in 'hard' systems, but spreads as a 'field' through the whole volume of the reaction mixture, acting at every point of the reaction network which is connected chemically to the reaction chain. This action chain is also error free, since of the order of 10^{20}–10^{22} molecules are involved in transporting the action.

The most important element in regulation technology is the closed action chain or, as it is more commonly called, feedback. The chemical cycle is no more than a closed action chain—a chemical feedback—as was discussed in a previous section.

In automatic devices the sign is usually weak and thus cannot be used to influence the function of machines and instruments directly. The weak sign must first be amplified. Amplification also has its chemical analogies—the autocatalytic processes discussed earlier and chain reactions (not to be confused with the simple reaction chains), which are similar to autocatalytic reactions. In fact, one group of action chains are acceleratory reactions; the mechanism of explosion and part of polymerization are both chain reactions.

The elements discussed above are already sufficient for the construction of rather complex chemical automata with complicated functions. For instance, a system capable of producing a time standard can be constructed, i.e. it can give signs depending on determined time periods. This would be a 'chemical clock' without visible hard fixtures—a solution which could be mixed and poured arbitrarily without disturbing its functioning. Indeed, it would function without errors, since each of its constituents is present in 10^{20}–10^{22} pieces, and thus the possibility of the failure of a part is excluded. Moreover, this chemical clock could be set as required to give the sign (e.g. a change of colour) by a determined number of seconds, minutes, hours, or even days.

Chemical automata have not yet been deliberately constructed because the theoretical principles have not yet been established. However, several chemical reaction systems of a clock-like character have been discovered by chance and reproduced experimentally. Such chemical systems are known as oscillating reactions. A vast literature on them is available, and they were the subject of popular and spectacular experiments in chemical lectures long ago. There are also many oscillating chemical systems in living organisms; these are the basic mechanisms of biological clocks.

The true automata are those that are controlled by a program. These require a program as well as information storage and decoding capacity. They have also been discussed earlier, although not from the present viewpoint of automata. However, we have demonstrated the existence of molecular signs, the existence of chemical information storage and decoding, and the molecular feasibility of transferring instructions by messages. We have seen that messages stored by molecular information can serve as programs in complex chemical systems; moreover, it is chemically possible not only to read but also to transcribe the program.

If chemical systems functioning according to the principles of regulation technology discussed above are combined with chemical systems corresponding to the principles of control engineering, then we have the possibility of constructing chemical automata with specific properties, such as a self-reproducing spherule or a chemoton. Beginnings like these predict a complete theory of chemical automata, which as it develops may transform the robotics of everyday life and produce an even greater revolution in automation than that produced by electronic automata following or from mechanical automata. However, all this lies in the future.

The chemical automata introduced above can do something (in addition to error-free functioning) which is just a dream in present-day robotics—they can reproduce themselves. Indeed, a self-reproducing spherule is already a self-reproducing chemical automaton and, as has been shown by computer simulations, the chemoton is a self-reproducing chemical automaton capable of producing self-reproducing automata more complex than itself. This has already been dealt with in detail, except that we discussed this feature of the chemoton in the language of biology and discussed its capability for evolution.

It seems that John von Neumann's dream of self-reproducing automata may come true in the not too distant future, but he could not have guessed that his self-reproducing automata would be chemical 'soft robots', and not electrical or mechanical devices.[G50]

3.14   Chemoton theory involves some basic laws of genetics

When I published in 1974 the ideas presented in this section in a scientific periodical under the title 'Theoretical deduction of the function and structure of the genetic material', the opinion was that it was easy to describe the structure of the genetic substance because of the discoveries made by Watson and Crick two decades earlier in 1953. Indeed, the structure of DNA was known and the laws of genetics were also known, but none of this involved chemoton theory.

Hence it is quite reasonable to discuss what is known from genetics in fact and what is implied by chemoton theory. What, if anything, can

[G50]John von Neumann's dream was discussed in his 1966 book (von Neumann and Burks 1966). von Neumann aimed to produce a consistent, rigorous logical description of a self-reproducing formal automaton, that is, a theory which would include in its scope both artificial automata like computers and natural automata like organisms. A key problem for such a unified, comprehensive theory is to understand how a formal description of a finite whole can be contained within it as a part to be transmitted to a new finite automaton in reproduction. It would seem that such a description would have to be as large as the system itself, yet if it were, either it could not contain it as a proper part or else reproduction must not proceed by transmission of a complete description upon which to base the developmental self-construction of the offspring. Biologists have generally opted for the latter conclusion in describing organisms: genes do not contain a complete description of the organism, so there is no paradox of formal description. The dream of a self-reproducing automaton is to discover the principles by which such a machine could in theory be constructed and thus achieve the reduction of biology to automata theory. See also Kemeny (1955) and Penrose (1959) for popular expositions of this dream.

chemoton theory add to the basic laws of heredity as revealed by classical genetics and molecular biology?

Inheritance in the strict sense means that the properties of the offspring are identical to, or are composed of, the properties of the parents. We have already seen that the concept of self-reproduction comprises inheritance in this strict sense (but does not, of course, comprise inheritable variability). In the living world the material basis of heredity is DNA (in some viruses it is RNA). This view is now almost universally accepted in biology. When John von Neumann outlined the conceptual construction of self-reproducing automata, he thought that three subsystems were necessary for this purpose, one of which was the subsystem giving the technological description of the self-reproducing automaton.

Nevertheless, it can be seen from our preceding discussions that self-reproduction does not require technological description or genetic material. Self-reproduction, and therefore inheritance in the strict sense, can be achieved without an information-carrying subsystem. Any molecule in an autocatalytic cycle can be considered as a molecular self-reproducing automaton. As it passes through the sequence of chemical states determined by the cycle, it reproduces itself without any technological or genetic prescription concerning either self-reproduction or the properties of the new molecule.

However, some tenders may feel that the self-reproduction of these molecules is not a true multiplication. Therefore, consider the self-reproducing spherules which, as we have seen, are metabolizing excitable homeostatic systems which divide into two equal spherules after functional growth, i.e. they reproduce themselves. The properties of the offspring are also identical with those of the progenitor; thus we can speak of inheritance, although no genetic material or genetic information is present.[G51]

Naturally, heredity is bound to genetic substances in the living world. However, the presence of genetic substance is not a prerequisite for the reproduction of offspring with the same properties—genetic substance is not indispensable for heredity. Similarly, a technological prescription is not a prerequisite for self-reproducing automata. Nevertheless, systems without genetic information do not have the capability of hereditary change. With respect to biology, this means that there is no possibility of evolution. As to automata theory, automata without technological prescription, although capable of self-reproduction, cannot produce automata more complex than themselves.

Hence in this case chemoton theory reveals more general laws, of which living beings and von Neumann automata are special cases which possess the capability of hereditary variability as well as inheritance.

Furthermore, chemoton theory implies that hereditary properties are not exclusively bound to genetic substance even in the present-day living world. In fact, every living being has some fundamental hereditary properties which do not depend on nucleic acids or any other macromolecules, but on compounds with a low molecular weight. In order to understand this, we must return to the role of enzymes in living systems.

[G51]This implication of Gánti's chemoton theory led to my reinterpretation of the significance of John Maynard Smith's analysis of units of evolution (Griesemer 2000c).

The individual steps of a living being's metabolism are catalysed by enzymes. In general, a chemical reaction in a living being can only occur if the enzyme catalysing it is in a state appropriate for functioning. According to molecular biology, genetic information refers to the structure and synthesis of these enzymes; the presence of hereditary properties depends on the presence of the appropriate enzymes.

Although the presence of an appropriate enzyme is essential in the present living world, it is not the only prerequisite for the corresponding chemical reaction. The other essential prerequisite is the presence of a substrate.

In biology the presence of substrate is usually taken for granted, since it either comes from the nutrients or is produced during metabolism. In fact, this is generally true, but clearly does not hold in the case of autocatalytic self-reproducing cycles and reaction networks since the substrate is produced by the autocatalytic system itself. Not only does the functioning of a self-reproducing cycle require the presence of every enzyme, but the process cannot start until at least one molecule of (any) one of the inner components of the self-reproducing cycle is present. On the other hand, as soon as one of the inner components is presence, the process not only starts, but works increasingly intensively, since the quantity of the substrate increases exponentially because of its self-reproducing character.

If, during division, a cell is produced without any of the components of its self-reproducing cycles, this cell will not possess the property determined by the cycle even if all corresponding parts of its genetic material are completely intact (unless the necessary component can be produced by an alternative metabolic pathway). This property will also be absent from the cell's descendants, i.e. the failure is hereditary.

This apparently curious statement does not contradict the general biological experience; it was only because of genetics and molecular biology that the inheritance-determining role of the genetic material has been rather one-sidedly emphasized. In fact, the above considerations do no more than suggest a mechanism for the observation that a living organism can never be developed from genetic material alone, i.e. only from chromosomes or the nucleus.

An egg, a seed, or a spore must always contain the substances of the cytoplasm. However, this hereditary factor depends on the environment. Let us examine this using a simple model, which is a slightly more complex version of our self-reproducing spherule. Let two cycles function within the same spherule, rather than one as assumed so far:

$$\mathbf{A} + \mathbf{X}^A \xrightarrow{\;A_1\;} 2\mathbf{A} + \mathbf{Y}^A + T \tag{3.44}$$

and

$$\mathbf{B} + \mathbf{X}^B \xrightarrow{\;B_1\;} 2\mathbf{B} + T + \mathbf{Y}^B. \tag{3.45}$$

Equation (3.44) is identical to eqn (3.27); the superscript A on the nutrient **X** and the waste material **Y** denotes that the nutrients and waste materials of the two processes are not necessarily identical. Equation (3.45) also describes the functioning of a cycle analogous to that described by eqns (3.44) and (3.27), but here the self-reproducing cycle passes through other kinds of inner components. Let both cycles produce the membrane-forming component T.

The functions of the spherule containing these two types of cycle is described by the equation:

$$\left[0.5n(\mathbf{A} + \mathbf{B}) + \boxed{T_n}\right] + 0.5n(\mathbf{X}^A + \mathbf{X}^B)$$
$$\xrightarrow[\;\;\underset{1}{\bigcirc}\;\;]{A,\,B} 2\left[0.5n(\mathbf{A} + \mathbf{B}) + \boxed{T_n}\right] + 0.5n(\mathbf{Y}^A + \mathbf{Y}^B) \qquad (3.46)$$

The ratio of the quantities of components **A** and **B** within the spherule may change; for instance, it may depend on the concentrations of both $\mathbf{X}^B$ and $\mathbf{X}^A$. However, the total quantities of **A** and **B** are fixed according to the rules already discussed.

It is well known that the rate of chemical reactions depends on temperature—it is faster at higher temperatures and slower at lower temperatures. The temperature dependence of different reactions is not the same; thus it is unlikely that cycles **A** and **B** will react identically to temperature change. Let us assume that at low temperatures the functioning of cycle **A** will slow down to more than that of cycle **B**. What will happen if a self-reproducing spherule containing two such cycles is put into a permanently cold environment? Obviously, the quantities of the components of cycle **B** will increase with respect to those of **A**. This also means that the self-reproducing spherule becomes more active in the sense that it functions more intensively than at its original composition, i.e. it adapts better to its environment. Under appropriate environmental conditions and a long adaptation period, the quantity of cycle **A** components may be reduced so much that eventually a division will occur in which not a single molecule of the components of cycle **A** will be passed into one of the two daughter cells. The function of this spherule is described by the equation

$$\left[n\mathbf{B} + \boxed{T_n}\right] + n\mathbf{X}^B \xrightarrow[\;\;\underset{1}{\bigcirc}\;\;]{B} 2\left[n\mathbf{B} + \boxed{T_n}\right] + n\mathbf{Y}^B \qquad (3.47)$$

Cycle **A** can never appear again in this spherule and its descendants even if the spherule is replaced in the warm environment. Thus our self-reproducing spherule has become permanently adapted by cold simply by being cultured within continuously cold circumstances. This hereditary change has nothing to do with the genetic substance, since the system does not contain a genetic prescription.[S33]

Strictly speaking, we have contradicted our previous considerations, since we have stated that the self-reproducing spherule is not capable of

[S33]Gánti demonstrates that, at least theoretically, holistic replicators can be units of selection because they multiply and are capable of heredity. Could they be units of evolution, i.e. show hereditary variability? In the case discussed this would amount to saying that, cycle A occasionally produces an intermediate of cycle B, analogous to mutation in the genetic material. Wächtershäuser (1988, 1992) believes that this is a valid option. However, we should bear in mind that up to now the only example of a non-enzymatic metabolic network is the formose system, which is incapable of heredity (there are no alternatives of the network that could propagate more or less accurately). Nevertheless, Gánti points to an interesting possibility: metabolism might evolve without a genetic substance coding for it. While the example is one of loss (of the A cycle from a system containing both an A and a B cycle, each sufficient for metabolism), it is also possible that a self-reproducing B spherule could acquire a new A cycle by accidental incorporation of one of its components from the environment. This would be a form of molecular symbiogenesis.

inheritable change, whereas inheritable changes and evolution require a genetic prescription. In fact, these non-macromolecular inheritable changes are extremely special cases of very limited importance; they are not mutational in character, they do not work in the direction of a more complex organization, and they occur very rarely. Only systems with macromolecular genetic material are capable of information storage and hence of inheritable changes which are mutational in character. Because of their enormous variability, such changes make evolution possible. Our results obtained from chemoton theory are of a more general character, which include the special cases of genetics and molecular biology.[G52]

Similarly, examination of inheritable changes of a mutational character leads to more general laws based on chemoton theory. Genetics and molecular biology have revealed in detail how inheritable mutative changes can occur via enzymatic mechanisms. However, as will be seen in the next section, the appearance of enzymes in living organisms must have been the result of a long evolutionary process and must have been preceded by a long and complicated evolution of systems without enzymes. Again, this would be impossible if inheritable changes and evolution could only be accomplished by changes in enzymes. In what follows, we shall show that mutative changes are general properties of systems with chemical informational subsystems, and mechanisms based upon changes in enzymes are merely special cases of these. However, these enzyme-based mechanisms are found generally, and apparently exclusively, in the present-day living world.

For simplicity let us ignore the function of the membrane subsystem and consider only the function and connections of the two other subsystems. Let us consider two distinct chemotons, whose functioning is described by the equations

$$[n\mathbf{A} + pV_n] + n\mathbf{X}^A \xrightarrow{\ A\ } 2[n\mathbf{A} + pV_n] + n\mathbf{X}^A \tag{3.48}$$

$$[m\mathbf{B} + pZ_m] + m\mathbf{X}^B \xrightarrow{\ B\ } 2[m\mathbf{B} + pZ_m] + m\mathbf{Y}^B. \tag{3.49}$$

Equation (3.49) is analogous to eqn (3.48) but, instead of a self-reproducing cycle built from components **A**, it contains another cycle, built from components **B**, which synthesizes a compound Z similar to but not identical with V. Let compound Z also be capable of polymerization according to an appropriate poly-Z template molecule (pZ_m). The length m of the template can be arbitrary; When $n = m$ it can even be equal to the length of pV_n occurring in eqn (3.48).

Let us now combine these two chemotons by joining the two types of template polymer at their ends:

$$\cdots V-V-V-V + Z-Z-Z\cdots \rightarrow \cdots V-V-V-V-Z-Z-Z\cdots$$

[G52]Gánti's view that molecular genetics describes a special case of phenomena described by chemoton theory inspired my development of the view that genetic and inheritance processes are special cases of reproduction processes (Griesemer, 2000a,b,c).

or

$$pV_n + pZ_m \rightarrow pV_nZ_m. \tag{3.50}$$

The combined chemoton now works as if it were the sum of the two originals. The compound V required for the reduplication of the (combined) template polymer is produced by cycle **A**, and compound Z is produced by cycle **B**:

$$[n\mathbf{A} + m\mathbf{B} + pV_nZ_m] + n\mathbf{X}^A + m\mathbf{X}^B$$

$$\xrightarrow[\;\;\;\;]{A,B} 2[n\mathbf{A} + m\mathbf{B} + pV_nZ_m] + n\mathbf{Y}^A + m\mathbf{Y}^B. \tag{3.51}$$

This equation is similar to eqn (3.46) which describes the functioning of the system capable of inheritable adaptation. However, the ratio of **A** to **B** is not arbitrary since it is determined in a hereditary manner by the ratio of V to Z in the template polymer pV_nZ_m.

Template polymerization is not a perfectly uniform process. Occasionally, an error may slip into the copying mechanism; for example, a Z molecule may be replaced by a V molecule. Then the number of V molecules in the polymer will increase by one and the number of Z molecules will decrease by one to give $pV_{n+1}Z_{m-1}$.

We have seen that the template polymer determines the function of the cycles in a hereditary manner. Hence the descendants produced after the error will be changed, and their hereditary composition will be described by the new expression

$$[(n + 1)\mathbf{A} + (m - 1)\mathbf{B} + pV_{n+1}Z_{m-1}].$$

If the same error is repeated in one of the following generations, the composition will become

$$[(n + 2)\mathbf{A} + (m - 2)\mathbf{B} + pV_{n+2}Z_{m-2}].$$

It is obvious that if the error is repeated m times during a long sequence of generations, the molecule Z will disappear completely from the polymer and hence cycle **B** will disappear from the system, i.e. we come back to a simple chemoton of the type described by eqn (3.48). If the errors occur in the opposite direction, i.e. if V molecules are exchanged for Z molecules, then cycle **A** will begin to disappear from the system and after a sufficient number of these transcriptional errors we return to the simple chemoton described by eqn (3.49).

Thus in this mutative chemoton model we have introduced basic mechanisms of inheritable changes which do not occur by gradual adaptation but as a result of random errors. The basis of the hereditary change is the same as in biological mutation—monomers building up the genetic substance are exchanged. In the language of molecular biology, 'base substitution' occurs. Nevertheless, whereas in mutations occurring in the present-day living world the change of information due to a base substitution is always indirectly manifested by the synthesis of enzymes, in chemoton theory it is also possible, in principle, via a direct chemical route. Before enzymatic mechanisms appeared, the living world may have passed through a sequence of mutations and a corresponding evolutionary route which eventually led to the development of enzymatic regulation.

Two fundamental conclusions can already be drawn from. First, in chemoton theory the laws of inheritance are dealt with on a more general basis than in molecular genetics, at the level of principles rather than particular substances. Hence chemoton theory may be capable of revealing mechanisms beyond the scope of genetics and molecular biology, which are based on the study of present-day living material. At the same time, the fact that the laws revealed by genetics and molecular biology are included within the more general laws implied by chemoton theory strengthens the correctness and biological validity of the latter.

In order to appreciate the second inference, we must remember that our considerations so far have not been based upon actual observations of biology or genetics. Everything has been deduced from a chemical, physical, and system-theoretical basis and the results have only been compared with experience taken from the present-day living world to show that our conjectures not unreasonable. This means that the results do not depend on the specific conditions of the terrestrial living world, our starting point does not contain *ab ovo* the end results, our deductions are not circular, and they are not pseudo-deductions. Moreover, our results hold not only for the present-day terrestrial living world, but also for earlier conditions and for extraterrestrial living organisms, if they exist, even if they are not constructed from nucleic acids and proteins. In order to see this let us deduce the broad outlines of the general theoretical molecular structure of the genetic material.

So far, we have not set many conditions on the structure and properties of the genetic material pV_n (or, in the mutative model, pV_nZ_m). All that we required was that it should be a linear polymer and possess template properties, i.e. it should serve as a pattern for its own construction. Let us now deduce by pure speculative reasoning, i.e. independently of the actual structure of nucleic acids, what kinds of structural properties are necessary on a molecular level to enable pV_n to satisfy these two conditions.

In order that the genetic message should be sufficiently stable, the individual signs, i.e. the molecules V, must be bound by sufficiently strong chemical bonds. Hence the chemical bonds along the strand must be strong covalent bonds. Furthermore, a linear polymer can serve as a template for freely moving molecules in the solution only if the monomeric molecules V can come into contact with the molecules V in the template molecule, and this also requires a chemical bond. However, this bond must be far weaker than the covalent bond in the chain since we have assumed that, after polymerization, the newly synthesized strand can break off from the template, i.e. bonds between the old and the new strands can be broken without breaking the bonds along the strands. The difference between the strengths of the two types of bond must be at least an order of magnitude.

Since the template governs the construction of the new strand according to monomeric units, the weak bonds must also be formed unit by unit (Fig. 3.20).

The longer the new polymer, the more weak bonds bind it to the old strand and hence the stronger the binding of the newly synthesized strand to the template. The binding is strongest when the new strand is exactly the same length as the template. Hence the double-strand state is the most stable of all the possible states. However, the template cannot have a single-strand structure because the newly synthesized strand cannot separate as it is strongly bound to the other. Thus the genetic material must have a double-stranded structure.

The molecular structure of DNA was deduced from the geometric structure of the individual monomers—the nucleotides. The chief concern of Watson and Crick was to find out what types of macromolecular structure can be constructed from nucleotides of given molecular configurations, and they found that one of these—the well-known double helix—corresponds exactly to the structure, function, and physical properties of DNA in cell nuclei. Thus they obtained their results for the structure of a specific molecule—DNA. It was realized later that RNA might have a similar structure.

However, we made no assumptions about the chemical configuration of monomer V. Starting from the most general functional property, namely the ability to form a template, we arrived directly at the double-stranded structure, which must also hold for the material storing chemical information in any self-reproducing soft system, not just DNA. Therefore we arrived at the same result from the opposite direction. We do not state that the double-stranded structure of the genetic material is a result of the steric configuration of the monomers. Rather, we state that the genetic material itself must have a double-stranded structure and thus can only be constructed from monomers which allow the formation of this structure.

We must add a further comment. The deduction of the double-stranded structure summarized above is necessary only if the genetic material occurs alone (in most of the present-day living world DNA is combined with proteins—nucleoproteins) and is synthesized by simple chemical means and not by enzymatic mechanisms. Otherwise, stability does not imply the double-stranded structure, as the binding to the template could also occur via strong bonds and the detachment of the newly synthesized strand could be promoted by enzymes. Thus the double-stranded structure of the present-day genetic material, as well as the presence of weak bonds in it, may indicate that its formation and its initial evolutionary phase preceded the appearance of enzymes.

Similar reasoning, which is not given in detail here, shows that there is only a negligibly small probability that the double-stranded structure would be straight, like a ladder, rather than helical. Similarly, the probability that the monomers themselves serve as templates (i.e. V for V,

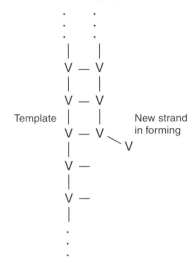

Fig. 3.20   Formation of new strand.

Fig. 3.21 Formation of monomer pairs.

Fig. 3.22 Structure capable of mutation.

Z for Z) is also very small; it is far more probable that complementary monomers are paired as templates for each other. In this case a monomer V must have a partner W so that V with W and W with V would form a pair in the chain (Fig. 3.21). In this case, however, the sign of the genetic material in the double helix is not represented by the molecule V but by the monomeric pair V–W.

A chemoton-like soft system with mutative capacity requires at least two kinds of signs, i.e. two kinds of monomeric couples. Hence we need a second pair Z–S in addition to the pair V–W.

Thus genetic material that is capable of mutation must consist of at least four kinds of monomers, arranged in a similar manner to that shown in Fig. 3.22.

We started from the laws of soft systems with non-enzymatic regulation and the ability to mutate and evolved, and deduced from this that the structure of the genetic material must be double stranded with complementary pairs of at least four types of budding monomers. This is exactly the structure of the DNA according to Watson and Crick, but we arrived at this result independently of the actual configuration of the monomeric building blocks.[S34]

Our laws for the structure and function of the genetic material are not restricted to DNA. In other words, if there are 'little green men' living elsewhere in the universe, the genetic information in their cells will be stored in this same general way, forming the basis of their mutability and evolution. The genetic material in their cells would be constructed from four kinds of monomers by complementary pair formation in a double-helix structure, even if this was be chemically very different from DNA.[S35] Thus the genetic laws implied by chemoton theory invade the realm of present-day living beings.

3.15   Chemoton theory involves an explication of the origin of life

Among the great problems of biology, the one with the longest history is undoubtedly the question of the origin of life. Both ancient religious texts and primitive myths all contain some idea of, or at least an allusion to, the origin of the world, the Earth, and life, showing how this question has excited the human imagination from the earliest times. The problem of origins also appears in early scientific writing. Epicurus, Aristotle, and Lucretius all dealt with the problem of biogenesis; moreover, they did not restrict the origin of life to Earth, but emphatically maintained that life is not unique to Earth.

In the eighteenth and nineteenth centuries, the question of the origin of life was restricted to life on Earth and many 'experiments' concerning the spontaneous origin of life were performed. The ideas and specifications underlying these experiments were primitive, and therefore the French Academy decided to conduct a competition to prove or refute experimentally the possibility of the spontaneous origin of life. The winner of the competition was Louis Pasteur with his famous experiments proving that abiogenesis does not exist. Every living organism descends from another one; every cell originates from another cell. And now we can add that every gene is produced by another gene.

Pasteur's experiments were exact and irreproachable, and so the question of abiogenesis was put aside for a while. At the same time, Aristotle's teachings about 'panspermia' were revived, and in the form of 'lithopanspermia' and 'radiopanspermia' diffused beyond scientific circles into the wider world. According to the doctrine of lithopanspermia, germs of life enclosed in meteorites wander about freely in space, while according to radiopanspermia they are carried by radiation. Both doctrines state that random encounters with celestial bodies appropriate for life would then provide opportunities for these germs to thrive and develop. Neither lithopanspermia nor radiopanspermia can be excluded with absolute certainty, but even if they occurred, they would not explain the ultimate origin of life.

In the 1920s Oparin in the Soviet Union and Haldane in Britain repeatedly emphasized the fact that life had not originated in conditions favourable for present-day living organisms. Conditions on the surface of the primeval Earth differed enormously from those of the present day, and therefore Pasteur's experimental results, although undoubtedly exact, do not exclude the possibility that life could have originated spontaneously under primeval conditions. Because of the efforts of Oparin and Haldane the problem was never again forgotten, and a series of comprehensive experimental investigations were started in about the middle of the twentieth century. The main lines followed by these experiments demonstrated with scientific exactness the first part of the complicated process, namely the formation of the basic materials of living matter—'chemical evolution'.

It is reasonable to pose the question of the origin of life in philosophy or ideology. However, the question is rather unreasonable in scientific terms since life itself is not exactly defined. In fact, in order to be able to explain the development of the living from the non-living, it is necessary to understand the difference between the living and non-living states. If this difference is not understood, the question cannot be answered.

Of course, there are many differences between living and non-living systems, but from our point of view the difference of principle is essential. For instance, living and non-living systems certainly have different material compositions: non-living systems generally consist of inorganic compounds, whereas living systems are composed of organic compounds. This difference raises the question of how organic compounds came into being. Now, this is a scientifically reasonable question, since we know the basic difference of principle between organic and inorganic compounds, and thus the origin of this difference can be investigated by exact scientific methods as has been done during the last three decades.

However, the difference between living and non-living is not identical to the difference between organic and inorganic. Therefore the difference between living and non-living must be established before we can begin to

[S34]The snag is that there could be *more* base pair types. In fact Benner's group (Piccirilli *et al.* 1990) designed alternative base pairs and managed to synthesize one of them, which can successfully be incorporated into nucleic acids. The question then is why there are no more base pairs in the genetic material. As Orgel (1990) remarked, there are two possibilities: first that evolution has never experimented with more than two, and second that for some reason two base pairs were found to be sufficient. This is currently known as the problem of size of the genetic alphabet (not be confused with genome size, which in the chemoton is the length n of pV_n). Szathmáry (1992, 1993) attempted an answer on the following lines. Suppose that protocells lived through a long RNA world era. Then RNA molecules served as genetic material as well as enzymes. Catalytic efficiency increases with the number of monomer types (active sites can be fine tuned more accurately), but copying fidelity decreases with it (monomers begin to resemble each other more closely and therefore the mutation rate rises). It can be shown that there must be an optimum number of base-pair types.

[S35]It is important that nucleic acids, and similar hypothetical structures, can form replicators with unlimited (indefinitely large) hereditary potential, in sharp contrast with holistic replicators (Maynard Smith and Szathmáry, 1999). Therefore the strongest prediction one can make is that evolved creatures in a foreign biota will possess a digital inheritance system. This is likely to have a double-stranded character, and it is not unlikely that there will be four letters in their genetic alphabet.

[S36]Knowledge of a minimal system does not by itself show how its subsystems came into being, and became assembled, in a *historical* sequence. The difference of view between my view (Szathmáry) and that of Gánti in the text depends on the kind of historical process by which living systems arose from non-living ones. Gánti's reference to a 'spontaneous' origin suggests a chemically deterministic process such that the origin of life might be deducible from chemoton theory plus an accurate description of the chemical starting conditions. Szathmáry's point implies that the historical sequence of steps, even if these are chemical steps, is not determined by the principles of chemoton theory alone, nor by chemoton theory plus starting conditions. The reason is that once the subsystems are on the road to assembly into chemotons, they change the conditions under which further assembly can proceed and the path by which previous subassemblies were reached may be obscured by the subsequent path of the process. This need not be due to chemical evolution, but to the broader effect of historical change in which the system itself plays a key role in shaping the historical context of its further prospects of change. For a philosophical discussion of historical explanation in biology, see, Lewontin (1978, 1983), Nitecki and Nitecki (1992), Wright *et al.* (1992), and Griesemer (1996a,b).

[S37]Trust in the 'primordial soup' scenario is no longer very great. First, the primordial atmosphere is unlikely to have been reducing; presumably, it was about neutral. *(contd.)*

study the question of the origin of life. Moreover, this question is not a philosophical one. We must give scientifically exact answers to questions such as: What is life? What kind of material, organizational, or other characteristics produce the qualitatively new properties appearing in living systems compared with non-living ones? Until we are able to explain exactly the principle of life, we cannot establish whether abiogenesis, i.e. the spontaneous origin of living from non-living systems, can occur. Only when we understand the principle of life can we hope to find a direct way of understanding its origin.

Of course, I believe that the chemoton theory correctly explains the fundamental principle of life. Indeed, if I did not believe this, I would not have written this book, which even in its title alludes to this problem. However, if the answer given by chemoton theory to the question concerning the principle of life is indeed correct, it must be possible to use it to deduce the spontaneous origin of living systems from non-living matter.[S36] In order to do this, we must establish the external conditions necessary for the spontaneous formation of living systems.

As has already been stated, according to Oparin and Haldane, the surface conditions of the primeval Earth differed substantially from those of the present day. The most important difference was not in the temperature or the amount of volcanic activity, but in the composition of the atmosphere: one of its main present-day constituents, molecular oxygen, was absent but it contained many hydrogen-rich compounds including the diatomic hydrogen molecule itself. The major components of the primeval atmosphere were hydrogen, nitrogen, ammonia, methane, and water vapour; smaller amounts of hydrogen cyanide, hydrogen sulphide, formaldehyde, ethane, helium, and argon were also present.

In 1953 Stanley Miller proved experimentally that electric discharges in the presence of an aqueous phase in a gaseous atmosphere containing such reducing components produce organic acids of biological importance, including amino acids. The surprising results of these experiments were so demonstrative and simple that scientists in many laboratories were inspired to carry out further research. This work led to the highly supported scientific picture of chemical evolution—the spontaneous formation of organic compounds of biological importance.[S37] Of course, here we can only summarize the main events of chemical evolution, without going into details.

Excluding helium and argon, which are inert elements, the most common chemical elements in the Universe are hydrogen, carbon, nitrogen, and oxygen, i.e. the biogenic elements from which the substances of living organisms are mostly constructed. In interstellar space these elements are found in the form of atoms, ions, molecules, or simple compounds such as water, methane, carbon dioxide, ammonia, cyanic acid, and formaldehyde.

In certain places in the Universe, where these molecules composed of three to five atoms concentrated in gaseous form (e.g. in the gaseous envelope of celestial bodies or adsorbed on the surface of cosmic dust particles), reactions may be initiated by cosmic and other types of radiation. As a result, more complicated molecules, containing five to fifteen or more atoms, are formed such as formic acid, methanol, ethanol, acetaldehyde, cyano-acetylene, ethane, propane, propionic aldehyde, etc., as well as some simple cyclic compounds such as benzene, toluene, pyrrole, etc. Radioastronomic analysis of cosmic dust and gas clouds, and chemical analysis of meteorites and samples of Moon rock brought back to Earth have identified the presence of about 60 of these compounds. Some macromolecules (unsaturated cyanopolymers containing hydrogen, nitrogen, and carbon) may even be found among these compounds.

Where temperature and pressure conditions allow the water to stabilize in liquid form, these compounds may react with water as well as with each other, forming a large number of organic compounds. Most of these compounds are included in the fundamental substances in the present-day living world: the essential amino acids, many organic acids and aldehydes, sugars and other carbohydrates, nucleic acid derivatives, porphyrins, etc. Furthermore, the cyanopolymers may be transformed by water into protein-like compounds—the proteinoids—which may condense into microscopic structures (membranes, spherules, and strands).

In summary, we have established that in those parts of the Universe where water can be found in a liquid state, as on primeval Earth, biologically important substances accumulate in large quantities and in many varieties—small molecular organic compounds, macromolecules, and microstructures (Fig. 3.23). This is much more than a hypothesis; its details have been proved in thousands of experiments.

Miller-type experiments perform rather poorly under neutral conditions (the yield of organic matter of biological significance is very modest). Second, chemical incompatibility precludes the *simultaneous* soup-based synthesis of particular chemicals that should have been present *together* for biogenesis. Third, the soup would have been very dilute (Shapiro 1986). There are two possible solutions: the primordial pizza scenario and the crêpes scenario. In the first, particularly advocated by Wächtershäuser (1988, 1992), organic matter forms on the surface of pyrite in hot hydrothermal environments, while the metabolic intermediates are always surface bound. In the crêpes scenario (von Kiedrowski 1996), organic matter formed (by whatever means) in the soup becomes absorbed onto mineral (e.g. clay) surfaces, and polymerization takes place on the rocks (Ferris *et al.* 1996). All scenarios may have some validity; they are not mutually exclusive.

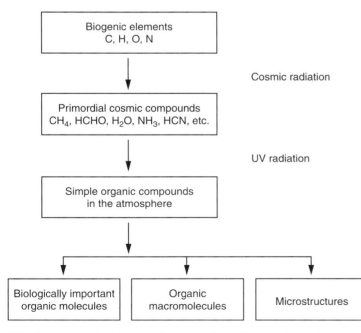

Fig. 3.23 Important steps in chemical evolution. The first three steps can occur anywhere in the Universe; the fourth step can only occur on the surface of celestial bodies where water is to be found in a liquid state.

However, an abundance of primeval water containing organic nutrients and constituents is not sufficient to initiate life. The set of all these compounds is not yet a living system. There is an enormous qualitative difference between even the most primitive living cells and the primeval nutrient medium containing the raw materials of living systems. The process bridging this gap has been called prebiotic evolution, since it is located between chemical evolution and biological evolution leading from primitive cells to man. Thus prebiotic evolution is actually the process of turning non-living into living—the process of becoming alive. It is at this point that present-day knowledge comes to a halt: there are no concrete experiments or exact theories about the process of becoming alive. This is not surprising, since one has to know what a living system actually is.

If we start from the hypothesis that the functioning of living systems depends on processes catalysed by enzymes with determined amino-acid sequences, we soon arrive at deadlock. Indeed, the probability of the random formation of a particular enzyme with specified amino acid sequences is vanishingly small. For example, according to Eigen's calculations, the 20 types of amino acids can be ordered in 10^{130} different ways in a protein containing 100 amino acids. In order to illustrate how unimaginably large this number is, Eigen calculated that the presently known Universe could contain, when closely packed, just 10^{103} of such protein molecules, so that a volume equivalent to 10^{27} universes would be necessary to accommodate all the possible variations of a protein molecule containing 100 amino acids.[G53]

[G53]See Eigen and Winkler-Oswatitsch (1996).

Thus it is clearly inconceivable that proteins of a pre-determined amino acid sequence have been formed by chance alone; there is neither enough time nor enough matter available for random formation. The formation of protein molecules requires guidance in the correct direction determining the amino acid sequence of the protein, depending on the properties of the substrate and of the genetic material. However, this requires the assumption that sterically confined self-reproducing systems containing both substrate and genetic material exist.

As we have seen, chemoton theory involves the assumption that the fundamental characteristics of living systems do not depend on an enzymatic regulation. Chemoton theory reveals organizational and functional laws of chemical soft systems working in a regulated and controlled manner, possessing genetic information, and capable of evolutionary progress, all without the necessity for enzymes. Hence chemoton theory avoids the problem by demonstrating that it is possible for living systems to form and function without any involvement of amino acid sequence of proteins.

Of course, this theoretical possibility does not mean that such systems actually exist. In nature, almost all biochemical reactions occur via enzymatic catalysis. Thus we have to prove that biologically important reactions or primeval spontaneous chemical processes may have occurred without enzymes. In principle, every reaction can take place without enzymes, since enzymes merely accelerate reactions, but this acceleration is of the order of 10^6–10^8 and thus biochemical processes without enzymes would proceed extremely slowly.

However, experiments on chemical evolution refute the latter assumption. It has been found that under reducing conditions corresponding to those in the primeval atmosphere, the formation of biologically important substances is not that slow at all. It was found that considerable quantities of biologically important substances are formed during periods of hours, days, or at most weeks. For instance, in Miller's experiments, discussed above, 15 per cent of the methane added was transformed into other organic compounds in 2 weeks, sugar-forming experiments performed by Ponnamperuma showed that 80 per cent of the formaldehyde introduced was transformed to sugar within 6 hours, and in experiments by Matthews and Moser 500 g of hydrogen cyanide produced within 2 weeks 200 g of cyano-polymer which was spontaneously transformed into proteinoids in the presence of water.

René Buvet and co-workers have recently performed intensive and successful experiments concerning the possibility of non-enzymatic metabolism. They pointed out that the formation of almost every type of biocompound, as well as the occurrence of biologically important reactions, may be possible at reasonable rates under conditions corresponding to those of the primeval atmosphere. Moreover, the possibility of catalytic acceleration in reactions is not excluded in the absence of enzymes. Many minerals, particularly some clays, have a catalytic effect on a number of reactions. Sidney Fox and co-workers have shown the catalytic effect of proteinoids (the spontaneously forming protein-like compounds) in reactions of a biological character.

Thus the chemical prerequisite of chemoton theory, namely the non-enzymatic realization of reactions of biological importance, has a real meaning with respect to the origin of life. If this is the case, in order to explain the process of becoming alive we must first investigate the possibility that chemical systems corresponding to the subsystems of the chemoton can be formed spontaneously under primeval conditions, and then see whether these can be self-organized into a single super-system—the chemoton.

Many types of reaction can occur under primeval conditions, as shown by experiments on chemical evolution in which a great many different organic compounds were produced. Other experiments prove that these compounds are not formed in a single chemical step but through a sequence of connected chemical reactions—reaction chains. The compounds formed in this way are chemically related and are the products of alternative pathways, so that the reaction chains are connected into reaction networks. To the best of my knowledge, direct experiments proving the existence of these reaction networks have not yet been performed. Nevertheless, all experiments which have attempted to reproduce primeval conditions suggest the formation of very complicated reaction networks in the reaction mixtures.

Experiments aimed at verifying chemoton theory or the process of becoming alive according to chemoton theory have not yet been performed. Neither have scientists investigating the events of chemical evolution attempted to find and locate cyclic processes. However, it is easy to envisage that chemical cycles must have had a major role under primeval conditions. The surface of primeval Earth was exposed to ultraviolet irradiation 10 times stronger than that of the present day, and absorption of this strong radiation by the different organic compounds produced energy-rich activated molecules. These activated compounds lost their 'superfluous' energy through a sequence of connected chemical reactions and when they returned to their ground state, they had completed a cycle.[S38] Under modern conditions, photosynthesis occurs by similar cyclic processes. This is quite understandable, as the continuous utilization of energy can only occur through cyclic processes.

Autocatalytic cycles are well known in experiments investigating chemical evolution. Buvet ascribes an important role to autocatalytic processes functioning in the primeval ocean. A good example of an autocatalytic process which has frequently been investigated (even from the viewpoint of technological utilization) is the formation of sugars from formaldehyde. This process passes through a sequence of many two-, three-, four-, five-, and six-atom carbohydrate molecules.

Thus we can concede that one of the three subsystems of the chemoton, the autocatalytic cyclic system, may have originated spontaneously; moreover, its formation and functioning in the primordial before the appearance of life was essential.

In the case of the second subsystem of the chemoton, the membrane subsystem, abiotic formation of two-dimensional fluid membranes was experimentally proved long ago. These experiments were not performed in order to prove chemoton theory, as they preceded its publication by almost a decade. The experiments were performed by an American professor, Sidney Fox, who prepared protein-like compounds from amino acids under the conditions of chemical evolution. Under appropriate experimental conditions, these proteinoids spontaneously formed double-layer membranes of electron-microscopic thickness which displayed the properties of two-dimensional fluids.[S39] Moreover, the membranes formed promptly closed up into spherules (as we assumed for spatially divided systems) so that the reaction mixture was filled with millions of identical spherules. The composition of the solution is different inside and outside the spherule. The identical spatial dimensions of the spherules suggest the existence of an optimal volume, which each of the spherules endeavours to achieve. Division of the spherules can be induced by changing the relation of the external and internal circumstances, i.e. by a physicochemical effect. Thus the membrane subsystem, including a mechanism for spherule formation and division, is not just theory; it can be realized experimentally under the conditions prevailing on Earth at the time of the origin of life.

Experimental proof of the spontaneous non-enzymatic formation of the third subsystem—information storage and reproduction—seems to be much more difficult. It is an intriguing question, since nobody has

[S38]Unfortunately, a cycle is not the only way of completing the system. An unwelcome alternative is the formation of chemically useless waste products—tars (Shapiro 1986). It is the balance between various possible reactions that matters.

[S39]The reference to proteinoid microspheres is of didactic value only. We are really interested in the spontaneous origin of lipid vesicles. Unfortunately, we do not have a plausible mechanism for the abiogenic formation of phospholipids. However, we do know that they can be delivered to the Earth in meteorites. The spontaneous assembly of vesicles from meteorite-delivered lipids has also been demonstrated. Ultimately, a plausible primordial pathway for lipid synthesis must be found (Deamer 1998).

succeeded in the experimental demonstration of the abiotic formation of nucleic acid macromolecules under the conditions of chemical evolution. However, this is not very important, since the synthesis of nucleic acids in the living world always occurs via a template mechanism. Thus we have to establish whether, under primeval conditions, nucleic acids could have been formed by a template mechanism without the intervention of enzymes.

Several experimental investigations of this problem have been reported. However, their results are not very illuminating. The process of polymerization was indeed greatly accelerated by the presence of template molecules, i.e. the template mechanism functions under primeval conditions without enzymatic catalysis, but it did not lead to true nucleic acid macromolecules—polymers consisting of a very large number of nucleotides. Non-enzymatic template polymerization of nucleotides always stopped at 'oligomers', i.e. chains consisting of only five or six nucleotides.[S40]

However, does not contradict the chemoton theory; in contrast, it seems to strengthen it. Problems would have been caused for chemoton theory by completely successful results, i.e. the formation of highly polymerized nucleic acids from nucleotide monomers through a template mechanism only, under the given experimental conditions.

We have established that the information-carrying molecule pV_n must have a double-stranded structure, since this is the only structure that is stable in solution. However, only a single strand can serve as a template, since in the double-stranded form the two strands cover each other's template surface. Now, in order to explain the results of the experiments on nucleic acid formation mentioned above, let us assume that a new strand with a length of more than four or five nucleotides can no longer break off from the template and thus, by blocking the template surface, hinders the synthesis of free nucleotide chains longer than five or six monomers. Indeed, it is well known from earlier investigations with a different aim that the maximum length of an oligonucleotide which can become spontaneously detached from the template at room temperature is four or five nucleotides, and thus polymers that are longer than this cannot be expected in chemical evolution experiments.

In the chemoton the double-stranded information-carrying molecule pV_n functions as a full template, making possible the synthesis of new pV_n molecules. How does this happen? In living organisms, the information-carrying substance, i.e. DNA, also has a double-stranded structure. This double strand becomes separated, forming two single strands which serve as templates for the synthesis (replication) of new DNA by complicated enzymatic mechanisms leading to separation. These mechanisms are not yet completely understood. As to the chemoton, we have seen that during its functioning the internal conditions are changed such that the separation of the double strand occurs via physicochemical effects only without the help of enzymes. It can be shown by rather circumstantial reasoning (not given here) that these conditions can only occur once in the life of each chemoton and necessarily lead to the synthesis of pV_n

[S40]Recently, polymers up to a length of 50 nucleotides have been obtained on a clay surface (Ferris *et al.* 1996).

molecules. As we have seen, the synthesis of pV_n starts the spatial division of the chemoton. Hence the rhythm-giving pacemaker role of the chemoton depends on the appropriate formation of these conditions for the separation of double-stranded pV_n. Without this pacemaker function, pV_n would be continuously synthesized in the chemoton such that it lost its internal temporal order. However, this internal temporal order fails in experiments concerning synthesis of abiotic nucleic acid by templates, and thus the absence of highly polymerized products indirectly proves our reasoning based on chemoton theory. Thus there is experimental support for the assumption that the informational subsystem of the chemoton may have originated and functioned under primeval non-enzymatic conditions.

Hence the abiotic formation and functioning of the three subsystems—autocatalytic reaction networks, two-dimensional fluid membranes, and the chemical informational system—are supported by exact experimental results, although these experiments were not deliberately performed to prove the existence and function of these chemical systems. Yet the results of these experiments are already steps in prebiological evolution, i.e. in the long process of becoming alive, which is no more than the self-organization of living systems. The appearance of a living system is just one more step from the existence and functioning of these steps. If the functioning of the three subsystems interlocks, a self-reproducing system capable of evolution, i.e. a chemoton, has been formed.

This last step has not yet been proved experimentally, as nobody has performed appropriate experiments.[S41] The success of such an experiment would be nothing less than the synthesis of a living system, i.e. the production of artificial life, the creation of life from non-living matter. The possibilities of this will be discussed in the next section.

3.16 Chemoton theory involves the strategy of the synthesis of living systems

At this point we should make a small digression. Man has often believed himself to be on the verge of finding the secret of life, so often imagined that only one more step has to be taken to complete the Grand Work as in the second part of Goethe's *Faust* where, after mixing and processing the appropriate material, the homunculus will finally be born in the crucible of the Great Magician! Are we not also looking for some scientific recipe for such a homunculus? Is there any recipe at all? Is it possible to produce a living being from non-living matter in the laboratory.

Many recipes for life have been suggested over the centuries. Let us cite just two from Oparin's book, both proposed by famous scientists. The first was suggested by Paracelsus, that fantastic mixture of scientist and magician who lived in the eleventh century. According to him, human sperm is put into a pumpkin, where a number of very complicated processes are performed on it. After a specific time, a tiny man is formed

[S41]Experiments by Luisi *et al.* (1994) have demonstrated that template polymerization could occur within reproducing lipid vesicles. However, there are two snags. First, the template is polymerized by an externally provided protein enzyme. Second, template polymerization is not synchronized with vesicle growth and fission. The incorporation of a metabolic cycle is a formidable problem, since the provision of suitably pure building blocks for template polymerization is not guaranteed (Chapter 4).

inside the pumpkin, with each part of his body just like a baby born from a mother except that each of its parts is much smaller. However, Paracelsus' recipe is very imperfect. The pumpkin is a living thing and the sperm also comes from a living being. Thus the recipe would not solve the synthesis of the living from non-living even if, miraculously, it could somehow be accomplished.

A somewhat better recipe is that proposed a century later by Van Helmont, the famous iatrochemist Brussels. It is described by Oparin as follows. According to Van Helmont human exhalations may serve as generating factors, and so one puts a dirty shirt into a pot containing some grains. After 21 days, the 'fermentation ceases and the exhalations of the shirt mixed with the exhalations of the grain will certainly produce some little mice'. This is a more artificial production of life, since it does not directly require living systems to take part, but nevertheless their 'exhalations' are necessary.

In the middle of the twentieth century a complete and, apart from a single error, scientifically irreproachable recipe for the production of living systems from lifeless matter was suggested by the famous English biologist, J.B.S. Haldane. His recipe is as follows: 'Take a planet with some carbon and oxygen, irradiate it with sunshine and cosmic rays and leave it for a hundred million years'.

Everyone who has read the previous section will agree that, except for the presence of oxygen, this recipe is perfect and is irreproachable from a scientific point of view. Indeed, it would certainly lead to the formation of living systems from lifeless substances. However, it would be somewhat difficult, to put it mildly, to perform the experiment under laboratory conditions. Nevertheless, up to now no better method for the artificial production of living systems has been suggested.

We must emphasize here, as we have done in the discussion of the question of the origin of life, that no better method could have been suggested. Indeed, if one wants to produce a living system artificially, the least one needs to know is what a living system is and what the fundamental characteristics of life are. Humanity dreamed of speaking machines thousands of years ago—just think of the Ark of the Covenant or of Hephaistos's speaking golden servants in Homer—but could even Leonardo have made a radio if the very nature and principles of broadcasting were unknown?

Living systems are incomparably more complicated than radios.

However, we must assume that the reproduction and artificial synthesis of living systems is simpler than the development of the radio. This statement is quite surprising, but could a radio have arisen spontaneously from materials found naturally under the influence of external circumstances only? Obviously, it could not under any conditions. Nevertheless, as implied in Haldane's recipe above, living systems can arise spontaneously, given the appropriate conditions. Under such conditions matter organizes itself spontaneously into living systems. Thus complicated devices do not have to be constructed for the synthesis of living systems; all that is necessary is that the appropriate conditions are provided, and the system will be self-organized. This spirit of the self-organizing universe may have been in Goethe's mind, when he wrote about the special importance of 'mixing' in the homunculus scene in *Faust*. But if this the case, why chemoton theory? And why has no-one yet synthesized a single germ of a living system?

The basic problem appears to be the failure of an adequate view. I strongly believe that in the present state of science every necessary condition is available for the artificial production of living systems. In principle, they could be synthesized in any well-equipped biochemical laboratory after the development of the necessary research methods and details, which should not take more than a few years. I am aware that this is not the general view of biologists. Indeed, I know of only two scientists performing experiments of this type; we shall return to their work later.[S42] Apart from them, no-one has yet tried to synthesize living systems—using modern scientific methods and this has a very simple explanation nobody knows where to begin. There is no theoretical basis for believing that experiments with this aim will be successful or are even reasonable.

This section aims to show that chemoton theory may provide a theoretical background for the artificial synthesis of living systems. On the basis of chemoton theory we can outline a plausible strategy for the artificial synthesis of living systems, as well as for the control of the living state of the systems obtained in experiments.

What kind of living systems should we try to produce experimentally? Which of the millions of plant and animal species should be chosen? Of course, the dream expressed in literature has been the production of a human-like creative which can speak and think—an intelligent homunculus. In the discussions of abiogenesis, mice, snakes, frogs, rats, worms, and flies 'were born' in such experiments. Spallanzani and, later, Pasteur had proved only the impossibility of the spontaneous production of moulds and bacteria. Obviously, the synthesis of living systems cannot begin with the production of mice and men. Moreover, the artificial *ab initio* synthesis of complicated living beings is not a question of inexperience and initiation. Presumably, the laboratory synthesis of higher multicellular organisms will never be feasible and the demand for it will never arise.

The laboratory synthesis of living systems is not what is envisaged by either the public or experts. Perhaps one day breeding fertilized eggs in a crucible will be a common procedure; perhaps human embryos will be brought to maturity under artificial circumstances from unfertilized eggs. The time may come when it will be possible to develop a complete animal not only from eggs but from any kind of somatic cell, and thus it will be possible to done thousands of identical living beings. We may even succeed in developing the large-scale industrial production of pork chops or beefsteak from tissue culture. However, all these techniques, given that there is a public demand for them, would just be manipulations of biological systems which already exist and would have nothing to do with the artificial synthesis of living systems. In fact, artificial production of

[S42]The situation is different now. Many laboratories have begun to perform preliminary experiments of this type. Gánti's statement that no-one tried the experiments because no-one knows where to begin has the ring of what ought to be truth, but rarely is in biology. Biologists seldom know where they ought to begin. Instead, they begin where others have left off, sometimes making abrupt turns, and then looking for the principles concerning where they ought to begin after they have more or less finished! This is one of the main reasons for the continuing relevance of Gánti's work, despite the fact that his principles were not needed to make progress in origin of life research. We need the principles now in order to develop theories that make sense of the many experiments and observations that were not motivated by knowledge of where to begin.

living beings presently found in nature would be quite unreasonable and aimless.

What is the aim of synthesizing artificial living systems at all? It will be expedient to begin our considerations by borrowing the concept of 'total synthesis' from organic chemistry.

Total synthesis is a process where a compound is produced not from other compounds but directly from its constituent elements, for example synthesizing a sexual hormone containing some 60–80 atoms directly from carbon, hydrogen, and oxygen. Of course, this is an extraordinarily complicated task for the chemist—hundreds and thousands of times more difficult and more expensive than producing the same compound from common substances with a similar chemical structure.[G54] Large-scale industrial production of more complicated compounds never occurs by total synthesis in the strict sense. What then is the use of total synthesis?

The precondition of exact chemistry is a knowledge of the molecular structure of different kinds of compounds. The approximately three million compounds of organic chemistry barely differ in their elementary composition; they are distinguished from each other by the organization of their constituent atoms. Obviously, organic chemistry could not exist without a knowledge of the structure of organic molecules. Many methods are used, mostly in combination, to reveal the structure. These methods can be used to construct exact structural models of molecules so small that they cannot be seen even under an electron microscope. However, after constructing the model it is still necessary to demonstrate its correctness experimentally; it must be shown that it corresponds to the molecular structure of the compound in question.

The best method for this purpose is total synthesis. Knowing the laws of chemistry, one can determine the individual steps of the synthesis necessary to obtain from the elementary components the final compound corresponding to the model. If the compound synthesized according to the model is identical with the original compound, then the model is correct.

Hence total synthesis has an indispensable role in the development of theoretical organic chemistry. In practice, however, it would be quite unreasonable to produce washing powders or insecticides by the method of total synthesis; much simpler, easier, and cheaper methods are available.

Similarly, the total synthesis of living systems, i.e. their production from non-living matter, has no other aim than proving their assumed organizational structure. All other aims can be achieved by simpler, easier, and cheaper natural methods, perfected by Nature over billions of years. However, every total synthesis requires a model for its completion. In the case of chemical total synthesis we have a model of molecular structure, and for the total synthesis of living systems we have an organizational model. This is where chemoton theory is of use in the artificial production of living systems as it is the only detailed organizational model of living systems established so far. The strategy of total synthesis can only be elaborated if one has an appropriate model,

[G54]Harold Morowitz once wrote an essay (1979, p. 3) about the 'Six million dollar man', to argue, *contra* a birthday card, that the economic value of a man was in millions of dollars, not pennies. The press calculated the cost of the elemental ingredients—hydrogen, oxygen, etc.—when purchased from a chemical supply house at about 97 cents. Morowitz calculated the cost of the biologically active ingredients—haemoglobin, ATP, etc.—when purchased from a biochemical supply house at around six million dollars.

since one has to know what compounds to synthesize and what kinds of organizational properties these compounds should have.

At this point, we must discuss the work of the two scientists mentioned above as the only researchers dealing with the question of synthesizing living systems. One of them is Sydney Fox, an American Professor, who was the first to make proteinoids by heating amino acids and then changing the conditions to produce microspheres. These experiments are of the utmost importance with respect to the research on chemical evolution and the origin of life.

Indeed, proteinoid microspheres have many extremely interesting properties. These can be classed into two groups. The first of these groups has the following properties: the microspheres are bounded by a two-dimensional fluid membrane, under given environmental conditions there is an optimal volume which each spherule attempts to reach, microspheres can be made to divide by changing the conditions, daughter spherules in a solution of membrane-forming proteinoids grow to the same volume as the parent; irradiation induces budding and in a proteinoid solution the buds grow, the membrane has selective permeability and the internal composition of the spherule differs from the external composition, etc. It is easy to see that all these properties correspond completely to the properties derived for the membrane subsystem of the chemoton.

In the second group, the properties of primitive enzyme activity and of associativity with nucleic acids can both be reduced to the macro-molecular characteristics of the membrane-forming basic proteinoids.

Recently, Professor Fox has began to consider these microspheres as living systems because of the properties discussed above. However, he does not give criteria for discriminating living from non-living systems. Our system of criteria is not satisfied by Fox's microspheres, since they do not metabolize (their growth is the result of the incorporation of ready-made proteinoids which are not produced by the system itself, and they have no information subsystem and consequently their processes are not self-controlled, but all events are regulated by external conditions). Nevertheless, the structural and functional properties of Fox's proteinoid microspheres correspond perfectly to one of the chemoton subsystems, namely the membrane subsystem. *Hence they can be considered as experimental proof of the artificial producibility and the spontaneous production of the membrane subsystem under primeval conditions.*

In fact, Fox's original aim was not to synthesize living systems, but to investigate the possibility of abiotic production of protein-like compounds. During his experiments he stumbled accidentally upon proteinoid microspheres—a proof in itself of the high probability of the formation of such systems—and only concluded that they could be considered as living systems after investigating them for a decade. However, his opinion was not based upon the laws of self-organizing living systems, but merely upon some 'biologicaloid' properties of the proteinoid microspheres.

The other scientist who has performed direct experiments involving immediate biogenesis is Krishna Bahadur from India. Bahadur and his wife, S. Ranganayaki, have been working on the spontaneous formation of microscopic spherules containing molybdenum atoms from primordial-like soups. Their experimental research has not attracted much attention in the scientific literature, although their results are the most interesting and most promising in this field. The reason for this might be that only modest experimental techniques and equipment are available to them, which are not in fashion today among scientists. Nevertheless, their spontaneously generated microscopic spherules—known as jeewanus from the Sanskrit for 'particle of life'—not only possess the ability to form membranes but also have some metabolic activity, including a primitive photosynthetic network. Thus these jeewanus appear to possess at least two of the subsystems of chemotons—the metabolic and the membrane subsystems. Bahadur and Ranganayaki have not investigated whether the jeewanus contain a primitive informational subsystem.[S71]

We can now outline a strategy for the total synthesis of living systems on the basis of chemoton theory. According to chemoton theory a living system must contain at least three subsystems: a self-reproducing cyclic system, an information subsystem, and a membrane subsystem. An appropriate combination of the functions of these subsystems automatically gives a living system. We have seen that experiments have already proved the possibility of both spontaneous formation and the artificial production of the individual subsystems. However, there remains a final step—combination of the three subsystems into a single functional super-system.

The synthesis of living systems depends on the possibility of finding three compatible subsystems which function when connected. We then have to find the system of connecting them, since it is far from certain that the autocatalytic cycle will produce the raw materials for the other two subsystems in the form required, or that it will produce all the raw materials needed for the other two subsystems. In the chemoton these connections are direct and are bound to one compound. The reason for this is that the chemoton is an abstract model of the simplest possible living system and it is highly improbable that it can actually be realized in this theoretically simplest form.

Therefore we have to show that it is not impossible to find three compatible (i.e. connectable) subsystems.[G55] It is reasonable to start from systems known to function experimentally, particularly since there are two which are distantly related. These are the autocatalytic sugar synthesis starting from formaldehyde and the template polymerization of nucleic acids. The former process produces carbohydrates containing three to seven carbon atoms from formaldehyde. One of these is ribose, a sugar containing five carbon atoms, which is one of the constituents of the nucleotide molecules serving as raw materials for the template polymerization of nucleic acids. The nucleotides have the structure

base – ribose – phosphate

[G55]In the light of notes S14, S38, and S41, connectability should not be equated with compatibility. Chemical compatibility requires, in addition to connectability, that there are no side-reactions which will hinder the functioning of the chemoton. Producing all the required raw materials as outputs of other parts of the system without harmful side reactions and without the use of highly specific enzymes is a problem that goes far beyond the connectability of the three subsystems. Moreover, chemical compatibility requires temporal synchrony of reactions (see note S41).

where the ribose and phosphate are the same in each of the nucleotides and the base is adenine, guanine, cytosine or uracil. Thus these nucleotides are connected through the ribose to the autocatalytic system with formaldehyde serving as the raw material. Thus through the autocatalytic cycle and the process of template polymerization may be connected functionally through ribose. In fact, these have been some recent reports of the spontaneous formation membranes, as well as microspheres, in solutions containing formaldehyde and ammonium cyanate.

Since one of the steps in nucleotide synthesis is the reaction of a sugar with phosphate, the autocatalytic cycle could also be based on sugar phosphates, too. In fact, the Calvin cycle functions through the transformation of sugar phosphates. Moreover, one of its inner components is ribose phosphate, the starting material of nucleotide synthesis in the present-day living world. Hence there is a natural connection between template polymerization and the autocatalytic cycle.

A sugar phosphates cycle is also suitable for forming a connection with the third subsystem, if the membrane consists of phospholipids and not of proteinoids. In fact, the basis of membranes in the living world is a matrix of double-layered phospholipids with the properties of a two-dimensional fluid, and phospholipids are synthesized acetyl coenzyme-A, which is formed in the transformation of sugar phosphates.

Thus three subsystems exist which theoretically can be joined into a single system. Moreover, the plant world functions through a combination of these same subsystems: plant cells incorporate the nutrients carbon dioxide and water through the Calvin cycle, nucleotides are synthesized from ribose phosphate formed in the Calvin cycle, and phospholipids are produced by the breakdown of sugar phosphates also formed in the Calvin cycle. Obviously, this is something like a chemoton of the plant cell.

However, it must not be forgotten that in plant cells all this occurs via the interaction of many enzymes through complicated enzymatic regulation. We have no empirical data about non-enzymatic functioning of the Calvin cycle under abiotic conditions. This has rarely been investigated, because simultaneous determination of the sugar phosphates in such a complicated system is one of the most difficult tasks in modern biochemistry. Nevertheless, the abiogenic formation of sugar phosphates and their transformation into other types of sugar phosphates is documented in the relevant literature.

However, the synthesis of living systems ought not to begin with the abiotic function of the Calvin cycle. Photosynthesis was a rather late development in the living world, and thus it is not certain that the Calvin cycle itself could be made to function appropriately under non-enzymatic conditions. Moreover, primitive cells had no need for photosynthesis, because the primeval soup contained plenty of raw materials produced by chemical evolution on which the first living systems could base their function by breakdown.

Now, three functionally connectable subsystems can be found based upon the breakdown of sugars. In this case the role of the autocatalytic cycle is played by the autocatalytic phosphorylation of adenine which, when combined with the breakdown of sugar, produces adenosine triphosphate (ATP). ATP is the general phosphorylating agent of living organisms. Both nucleic acid template polymerization and the production of membrane-forming phospholipids require organically bound phosphate, which is provided by the general ATP conveyor—autocatalytic phosphorylation.

Thus we have outlined several theoretical methods which could be used, in principle, to synthesize living systems, although it is unlikely that these actual systems will be combined in the first artificially produced living systems.

We have achieved the aim of this section: we have demonstrated the possibility of artificially synthesizing living systems or, if one wishes, the possibility of 'total synthesis'. Is this really the whole story? Was it not much ado about nothing? I think that the reader may be rather disappointed. After all, we generally expect the name 'artificial living systems' to describe some kind of homunculus.

Nevertheless, let us just remember the artificial release of atomic energy. Once its theoretical possibility was demonstrated, the first artificial transmutation of atoms was accomplished within two weeks(!) by Frederic and Irene Joliot-Curie. However, these laboratory experiments were not accompanied by terrible explosions, enormous heat, and other spectacular phenomena, but just the detection of a few phosphorus atoms by subtle methods. However, the scientists were well aware of the unimaginable prospects opened by these experiments. Indeed, barely a decade had passed before the first atomic reactor was in operation.

3.17   Chemoton theory involves the possibility of an exact theoretical biology

In the previous sections we have always been concerned with life. However, our treatment was not a full empirical biology or even a theoretical biology. The preceding considerations—the questions concerning the origin of life and the construction of artificial living systems—were nothing more than foundations for the development of an exact theoretical biology. Indeed, what is biology? It is a science dealing with living beings and with the living world. In our previous considerations, systems actually found in the living world have mostly been used as examples. However, we hope that our preceding deductions will serve as the basis for developing mathematically exact models which are suitable for describing more complicated systems of the living world. Thus actual biological phenomena can be treated with appropriate precision. Our aim in this section is to investigate the possibility of doing this.

However, at present this is only a possibility. Theoretical biology is not yet an autonomous science like theoretical physics, and it cannot be developed by a single scientist. Even at this level our considerations required the application of many disciplines (physics, chemistry, thermodynamics, cybernetics, etc.) and the knowledge of all these clearly surpasses the capacity of one person. Although the results presented here, except for the computer simulations, reflect the author's own investigations and rest upon more or less exact deductions, this was only possible at the price of considering only the fundamentals and ignoring the development of individual topics. For instance, the author did not wish to develop a cybernetic theory of soft automata or the irreversible thermodynamics of the stability of accumulation systems. This is not and cannot be our task; if it is necessary, it should be done by competent students of the respective disciplines.

In this section, we shall consider problems which have not yet been solved. Some solutions, for instance the principles governing the function of certain soft learning systems, have already been outlined; in others, such as a model of eukaryotes, the approach is tentative and requires further work. Finally, there are topics in which there are only indications of a hopeful solution. Since our final aim is the development of a theoretical biology, all the above have a place since some of them may stimulate a researcher, thus contributing to the development of theoretical biology.

Undoubtedly, the description of the present-day living world must contain enzymatic regulation. Enzymes have not occurred in chemoton theory and therefore chemotons have never been called cells. In fact, we have avoided using the terms 'primeval cells' or 'protocells' in connection with chemotons. However, a chemoton including enzymatic regulation certainly could be considered as a model of the cell. How can enzymatic regulation be built in into the chemoton?

It was mentioned in the discussion of information-carrying subsystems that information can be stored by the quantity of signs, by the quantitative relation of two or more signs, and by the sequence of signs. In the simple chemoton, information is stored by the quantity of signs; in the mutative chemoton it is stored by the ratio of two (or four) kinds of signs.

Although the sequence of signs played no role in the mutative chemoton, owing to the mechanism of template polymerization the sign sequence of the offspring molecule is identical with that of the template molecule. Hence the offspring of a mutative chemoton contain the same sign sequence and the same information, if the sign sequence carries information. *Thus the mutative chemoton has a capability as yet unused, i.e. to store information by sign sequences, which is of infinite capacity.*

It is easy to see that this is the point where enzymatic regulation can be introduced. The mutative chemoton provides a free capacity, which can be filled. This capacity is filled in nature, because enzymatic regulation depends on the sign sequence of the genetic substance; the sign sequences of DNA determine the enzymatic functions. The exact model of this function has yet to be constructed and must be integrated into the chemoton model in order to create a complete exact model of the living cell. We have already succeeded in incorporating enzymes into the stoichiometric equations of prokaryotes.

The integration of enzymatic regulation is not a preliminary condition for the construction of models of more complicated biological systems.

Let us emphasize again that enzymes cause changes in reaction rates, but properties depend ultimately on the occurrence or non-occurrence of chemical reactions in the organisms.

On our way towards more complicated biological systems it will prove expedient to return to our consideration of total synthesis, and to discuss again our statement that there will never be a demand for the artificial production of highly complicated living beings from lifeless matter. We have stated that, in chemistry, total synthesis means the direct synthesis of a compound from its elementary constituents. In principle this concept is still true, but in practice the meaning underwent a slight change of interpretation. In fact, in order to perform a total synthesis it is not always necessary to start from the elements themselves; in most cases it is sufficient to start from simpler compounds whose structure has already been determined in total syntheses performed by others. Thus if one wishes to synthesize cellulose in order to prove its structure, it is not necessary to produce it directly from carbon, hydrogen, and oxygen; it is sufficient to start from glucose, since cellulose is built up from glucose molecules and the total synthesis of glucose has already been carried out by other workers. Just as in the proof of a geometrical theorem, it is not always necessary to start from the axioms; it is generally sufficient to refer to theorem which has already been proved.

Complex biological systems are organized from simpler systems into systems with a higher degree of organizational hierarchy. The eukaryotic cell is organized from a few types of prokaryotic cell, multicellular organisms consist of millions of many types of eukaryotic cells, and animal communities, say colonies of social insects, are organized from various multicellular individuals into a higher system.

Hence if one attempts to solve the 'total synthesis' of a system on the organizational level of eukaryotes, it is not necessary to start from lifeless material. It is sufficient to begin with prokaryotes, provided that the synthesis of prokaryotes from lifeless substances has already been performed. Thus, as long as the synthesis of systems on a prokaryotic level is unsolved, eukaryotes cannot be 'totally synthesized'; however, if the total synthesis of prokaryotes is already feasible, it is superfluous on the level of eukaryotes since it is an intermediate step. Similarly, if one wishes to produce a multicellular organism, it is not necessary to start from chemical elements but from eukaryotic cells. However, this is already feasible, since complete plants or animals can be developed from a single somatic cell.

Now we can understand better the categorical statement that the demand for producing multicellular organisms, animals, or man from non-living matter will never arise. In fact, only systems on a prokaryotic level can be produced from lifeless substances; eukaryotes can only be produced from prokaryotes, i.e. from already living systems, and similarly multicellular organisms can only be produced from eukaryotic cells, i.e. from living systems built up from other living systems. Thus all these 'syntheses' are biological operations and not constructions from

G56Herbert Simon (1962) argued that this kind of 'biological' construction, i.e. construction of a whole system from stable subassemblies rather than from elemental constituents, is the subject of a very general 'science of the artificial'. In general, complex systems are more likely to be constructed from stable subassemblies than by strict total synthesis.

lifeless substances, since the prokaryotes obviously belong to the realm of biology.G56

If we wish to extend chemoton theory to more complicated biological systems, it will be reasonable to follow the same path. The chemoton is a general abstract model including the bases of every function of a prokaryotic cell except enzymatic regulation. Because of this abstract generality, it is suitable for generating many kinds of concrete chemoton models. Thus, restricting ourselves to examples from the known living world, the abstract chemoton can be constructed from models in which the self-reproducing cycle is driven by the fermentative decomposition of complete organic molecules, or is capable of incorporating carbon dioxide by using the energy of light, or transforms organic compounds into carbon dioxide and water with the help of atmospheric oxygen. Moreover, a chemoton model can be constructed using the energy of the nutrients not only for chemical work but also for mechanical motion.

It is easy to see that these chemotons are none other than models of different kinds of prokaryotic cells, capable of different kinds of metabolism (fermentative, oxidative, photosynthetic) and movement. We have seen that eukaryotic cells originated from a symbiotic connection of several prokaryotic cells. Thus in order to obtain the exact model of the eukaryotic cell, we have to combine different types of prokaryotic models into a single system in the same way as the three independent subsystems were organized into a single 'super-system'—the chemoton.

The path from eukaryotes to multicellular organisms can best be modelled in two separate steps: the development of models from different organs, and the unification of these models into a model of a single organism. As an introduction, it will be expedient to construct a model of a 'simple' organ. Let this be the brain.

Now, I do not wish to make bad jokes. I am perfectly aware that the brain is generally considered to be the most complicated organ in a living being, and with good reason. Indeed, the brain is capable of producing something, namely thinking, which is quite exceptional in the living world and whose mechanism cannot yet be explained. However, from the viewpoint of functional (or 'software') modelling, things are rather different. For example, let us compare the tasks performed by the brain with those performed by the skin or the kidney. How many influences act on the skin? It has to react to cold, heat, pressure, radiation, moisture, atmospheric composition, etc. It has to respire, perspire, defend against infection, regenerate, and perform many other functions. Similarly, the kidneys have to regulate the excretion of many compounds and maintain ionic concentrations in the blood.

What about the brain? It is enclosed in a hard shell and protected from external injury. The cortex does not have to react to direct effects from the exterior world, it does not have to defend itself, and, once developed, its neurons need not even multiply. Undoubtedly, it has a highly complicated and subtly organized structure. However, this intricate 'hardware', is far from being 'hyperdetermined'; there are large areas which can, with some exaggeration, be 'slashed' without fatal consequences. Indeed, the brain has a single task: to receive, process, and emit electric signs running along the nerve fibres. However, this 'single' task is far from simple. To be more exact, this is not just a task but a whole spectrum of tasks within which the brain has to solve a very large number of even newer tasks. There are so many kinds of tasks that it is simply impossible to imagine a separate mechanism for each. However, this fact may be a clue to the solution: the principle governing the functioning of the brain must not be sought in an information-storing substance, but in the infinite variability of the signs.

In principle, this is the reason why the functional model—the 'software'—of the brain cortex can be deduced from the chemoton theory. Only the neurons have to react to incoming electric signs, and thus a special chemoton model capable of this single function of the neurons can be designed. Moreover, appropriate variations of this basic model can be devised and, similarly to the construction of the chemoton where the combination of three subsystems led automatically to a super-system with surprising new properties, the combination of different types of 'neuronal chemoton' also leads to new qualitative characteristics which may be considered models of learning and perhaps even of thinking.

First, consider learning. During learning, the system becomes transformed and changed. Then the things learned begin to fade; the system forgets. However, it does not forget anything completely. It generally learns through multiple iterations, but also through single strong 'imprintings' which are never deleted. The system is even capable of 'dreaming'. Moreover, if it is confronted with insoluble situations, it may behave like something that has 'gone mad'; it may have 'fixed ideas'. In some cases these 'obsessions' may be 'cured', overtime, by electro-shock treatment. The functioning of the system becomes increasingly rigid, moving towards something like 'senility'. In this state more and more experiences collected in its 'youth' emerge, but in a quite unconnected manner. However, in crisis situations the system is capable of recalling everything learned in its 'life'.

These and similar properties follow directly from the model. The details cannot be given here, but the reader can find them in chapter 5 of Gánti's monograph, *The Chemoton Theory*, Volume 1. Of course, it is far from certain that the brain really works by a mechanism like this. Yet even in this case, chemoton theory has provided us with a detailed model of soft automata that can learn.

So far, not even the first steps have been taken towards constructing models of multicellular organisms from models of individual organs, but the way seems to be open since the 'recipe' has been given; we only have to find the appropriate 'demand and supply' coupling between the models of the organs.

Thus we have arrived at the end of our journey. Moving from molecules to multicellular organisms, we have outlined a possible exact theoretical biology. One may and ought to go further towards biocenoses and the whole living world, but this is a task for the future. The author has one more task left in this book: after following the path of scientific reasoning so far, he must now turn to responsibility and social insight. Let the last chapter belong to them.

3.18   The responsibility of the biologist

The history of science and technology appears to suggest that the progress of humanity and the development of science are accompanied by human tragedies. The victims of Arctic research and of explorations in remote and dangerous parts of the Earth are commemorated by books and monuments. A monument in the garden of the Saint George Hospital, Hamburg, bears the names of more than 200 victims of X-radiation and radioactivity. It would be virtually impossible to collect the names of victims of the development of chemistry—the many chemists killed by explosions, poisons, burns, and the corrosive effects of acids and strong caustics. A whole book could be written on the victims of aviation, and who could count the many victims of research and development in electricity?

In this book we have outlined topics such as the synthesis of living systems, the production of soft automata, and the development of soft systems which can learn. So far, topics like these have generally been dealt with by science-fiction writers. Moreover, we have treated them not as mere theoretical possibilities, but outlined a method by which they could actually be realized. Where will this lead? Can we follow this path without penalties? Is it not possible that catastrophe lies in wait? Would it not be better to stop now, or must we follow this path to its end? Doubts like these may arise in a thoughtful reader, as indeed they have arisen in the author. The process of biological manipulation did not begin with chemoton theory. Man has been led along this path by the domestication of animals, the cultivation of plants, and the application of the first drugs, but of course not consciously. The eventual harm was restricted and the damage was always dwarfed by the advantages. In fact, the damage was mostly not perceived at all, or at least its cause was not even suspected. But today biological manipulation has assumed enormous dimensions in almost every part of biology, from birth control to the breakdown of the environmental equilibrium. There is no need to dwell on these matters here, as they are dealt with in widely read books.

However, in the dangers evoked by biological manipulations there is something more threatening than, and qualitatively different from, the perils produced so far by technology and civilization. I believe that the qualitative difference is caused by two factors—irreversibility and the time factor. Let me explain this with reference to the story of the chemoton theory.

The first ideas for the chemoton theory arose in 1952 from the insight that every living system has two fundamentally opposite properties. One is that living beings must always compensate for changes of the environment in order to preserve their state of equilibrium; the other is that these same living beings must go through irreversible sequences of events which never recur, such as the processes of embryonic development, reproduction, or ageing. In other words, one of the processes must be bound (not necessarily in a thermodynamic sense) to reversible events and the other must be bound (again not thermodynamically) to irreversible events. Thus the two processes are mutually exclusive and must be realized by two subsystems functioning with two different mechanisms. In my book *Revolution in the Research of Life* (published in Hungarian in 1966), I called the first system the 'equilibrium subsystem' and the second system the 'chief cycle'. Later, these systems were renamed the 'homeostatic self-reproducing cyclic system' and the 'information subsystem' respectively.

Natural or artificial effects on the homeostatic subsystems of living systems, within the limits of compensability, do not cause lasting changes in it. However, direct or indirect effects on the information subsystem cause irreversible changes because of the irreversible functioning of this subsystem.

Undoubtedly, the concept of back-mutation is familiar from genetics; sometimes a mutational change becomes 'back-mutated'. Nevertheless, because to its low probability, this event occurs in a solitary individual after many generations and this cannot nullify the fact that meanwhile offspring of the original mutant will multiply and interact, eventually innumerably, with other members of the living world.

Since the basic units of the living world carry unidirectional irreversible sequences of events, the systems constructed from them will also carry unidirectional programs in their respectively higher levels of organizational hierarchy. Thus mitosis in eukaryotic cells, ontogenesis, i.e. embryonic growth and differentiation, in multicellular organisms, and phylogenesis at the level of the living world, are all instances of such irreversibility.[S43]

Here we arrive at a fundamental difference in principle between the living and the non-living world—the state of non-living systems is entirely determined by external circumstances. This was the fundamental insight of thermodynamics, leading to the concept of state variables. If the state variables (pressure, temperature, volume, etc.) are known, the state of a non-living system (except for a few cases) can be uniquely determined.

However, this does not hold for the living world. Events in the living world are affected not only by state variables, which naturally restrict the realm of possibilities, but also by other limiting factors. Such limiting factors specific to living systems are not only changes in the information subsystems of an individual and its progenitors, but also changes in the information subsystems of living systems due to mutual interaction with the individual or its progenitors. In summary, the living state is determined by external circumstances as well as by its own previous history.[G57] If an experiment of cosmic dimensions could reproduce exactly the development of Earth, the evolution of life in this experiment would not follow the actual course of phylogenesis exactly, since random events have an essential role in changes in information subsystems and

[S43]It is not clear to what extent phylogenesis is unidirectional or irreversible. For example, the Cambrian 'explosion' might have resulted in the net extinction of phyla.

[G57]This theme will be familiar to philosophers of biology through the works of Richard Lewontin (1978, 1983).

every random event, once it has occurred, will manifest itself unavoidably in the infinite manifold of progenitors.

Our characteristics are not just determined by those of our father or by how our grandfather had chosen his wife, but in a certain sense by things which occurred to some amoebae or blue-green algae three billion years ago. Moreover, these changes are not just due to the information written in their genetic material, but are affected by the genetic substance of other living organisms. For example, the characteristics of the reindeer are affected by what is written into the genetic substance of lichens or of the wolf.

This complex sequence of events is the essence of the irreversibility factor. Such sequences are extremely rarely, and peripherally, found in the non-living world, but in the living world they are a fundamental determining factor. It is this factor of irreversibility which makes biological manipulation incomparably more dangerous than manipulations performed on non-living systems. One can also produce catastrophes in the non-living world, but sooner or later every bad thing comes to an end. However, in the living world nothing really comes to an end, neither good nor bad. Every event which has an impact upon the irreversible processes will continue to exercise influence until the end of the living world. Thus the effect of these events is in practice endless.

The other factor is less fundamental but nonetheless hides dangers of similar importance. In the non-living world reaction generally follows action immediately or within a short time. Thus, if we strike a match in ethereal vapours they will explode immediately, and similarly if a bridge is badly overloaded it will collapse immediately. Thus individual catastrophes may be unavoidable but their repetition can be avoided, since we are aware of the immediate reaction. However, in biology even an immediate effect occurs after a relatively long delay. Even simple infection or poisoning has a latency in the organism of hours, days, or even months. Damage to the genetic substance will be manifested at the earliest in the next generation, which in the case of man means decades.

Indirect effects already have an evolutionary character and are manifested only after hundreds or even thousands of years. Thus we have no knowledge of them at all. Almost every one of our present-day actions produces an effect of some kind, which will accompany the evolution of the living world through millions of years as an inheritance, and unfortunately as a damned inheritance.

For most the long history of the species man has considered only the present; the past has been appreciated a little more during the short history of civilization. As to the future, until the present century this meant at most some decades, and consideration of this was left to the rulers.

But now we live in a different age. Our activities blindly and brutally affect the future, and this future no longer means a few decades but the whole future—not only the future of man but that of the entire living world. Today man's activities affect the future much more than the present. Whether we wish it or not, we influence the irreversible processes of the entire biosphere and we do so in ignorance. However, this interference cannot be prevented; it has its roots in the development of technology and civilization, and, just like the evolution of life, the progress of society is irreversible. The irreversibility of man involves the irreversibility of society, just as the irreversibility of the information subsystem involves that of the chemoton, then the prokaryotes, and then the eukaryotes, and the irreversibility of the eukaryotic cell implies that of the multicellular organisms which implies the irreversibility of the entire living world, which we call evolution.

There is no alternative. Human beings and human society cannot exist without the progress of science and technology. In a sense, 'manipulation' is unavoidable. Nevertheless, interference and manipulation need not necessarily imply catastrophe. Although the history of science suggests that progress has often occurred at the price of setbacks, accidents, and catastrophes, this is by no means a hard-and-fast rule. We have an actual example that catastrophe can be avoided, at least in scientific research.

This example is the liberation of atomic energy in the nuclear reactor. Needless to say, there was far greater danger associated with the development of nuclear reactors than in any previous technical advance, but no catastrophic errors occurred. This was not surprising, because an exact theory was first worked out which was capable of determining in advance events and necessary conditions with a qualitative and quantitative precision, and only then were experiments performed. Indeed, at the time of practical applications atomic physics was already a well-established and well-proved theory, supported by quantum mechanics and statistical mechanics.

Biology is still a mainly experimental and descriptive science. It has no theory which can be used for the qualitative and quantitative description of biological processes, the reactions of cells, multicellular organisms, biocenoses, or external or internal effects on the entire biosphere. Undoubtedly, there are some fields of biology where fruitful theoretical models for important partial processes have already been developed, such as the Watson–Crick model of inheritance or the various models of intracellular regulation. However, none of these are able to give a quantitative description and refer only to partial processes. We must know the responses of entire living beings and of entire biocenoses in order to understand the consequences of our interference.

Hence the creation of an exact theoretical biology is a crucial task. This must be an exact biological model system with exactly determinable properties so that it can approximate actual biological phenomena with the necessary accuracy.

In the case of liberating atomic energy—the first in the history of science—enormous sums of money were required and hence there was no latitude for trial and error; one had to be sure of the results. Neither worry about the future nor fear of catastrophe was enough to put theory and research before practice.

Unfortunately, in the case of biology the situation is the reverse. Here, the large sums are not required by the biological manipulations themselves, but very often, as in the case of environmental pollution, to prevent them. Thus the development of theoretical biology is not driven by direct material interests; the driving force must be human responsibility.

Indeed, the development of an exact theoretical biology is the precondition for the appropriate design of biological processes. This planning is important in the light of environmental policy, public health, demography, provisional policy, and scientific research alike. The possibilities for developing a potent theoretical biology are provided today. However, doing it effectively does not depend only on the biologists. Organizers of scientific research and governments share in this responsibility.

4 The biological significance of Gánti's work in 1971 and today

Eörs Szathmáry

4.1 *Annus mirabilis* 1971: Eigen and Gánti

Gánti's book was completed in 1968 and published in Hungarian in 1971. It was the era of triumhant molecular biology—by that time the basics of a bacterial cell, including the genetic code, were known. Everything revolved around proteins (mainly as enzymes) and DNA (the genetic material). Interesting steps towards the understanding of life's origin had been made—several organic molecules had been synthesized in Miller-type 'primordial' experiments, by Miller, Orgel, Fox, and others. Yet the connection between the two worlds of the molecular foundations of contemporary cell heredity and of 'primordial alchemy' was not established. Although people believed in the existence of such a connection, it was hard to see how life, including the genetic code, could have arisen. Some argued that one could synthesize life without the genetic code. In fact, Fox (1965) claimed that his proteinoid microspheres were alive. Even if one accepted this (which people did not), one still found the origin of contemporary organisms (such as bacteria) baffling, essentially because there was a lack of understanding of the origin of the genetic code.

Two major theoretical works pertinent to these questions were published in 1971: one was Gánti's book and the other was a long paper by Manfred Eigen which appeared in *Naturwissenschaften*. Both authors strongly emphasized autocatalysis (the chemical process in which one compound catalyses its own formation), and both failed to see the importance of compartmentalization for function and evolution. However, these are the only elements shared by the two hypotheses. Let us first explain Eigen's work in some detail because it provides a nice contrast to Gánti's ideas.

Eigen first realized (and this is a genuine discovery) that primitive information carriers composed of nucleic acids could not have been longer than a modern transfer RNA (tRNA) molecule because of the excessive mutational load arising from highly inaccurate replication in the absence of evolved enzymes. However, several unlinked genes would have competed in the primordial soup. This is why he invoked the hypercycle—a system of molecular mutualists in which genes were coding for specific replicases which in turn catalysed the replication of the next member gene in the cycle. The hypercycle rests on second-order autocatalysis: replication is catalysed by the templates as well as by the replicases. It can be shown that such a system is ecologically stable, i.e.

despite internal competition, stable dynamic coexistence of all the members is ensured provided that mutations do not occur. But mutations do occur, and this was found to lead to unwanted complications (see later). Eigen failed to provide a convincing scenario for the origin of the genetic code; in 1971 he merely postulated its existence. Therefore the origin of the system remained a mystery.

Gánti, in contrast, liberated himself from the burden of the genetic code. He argued that, although enzymes are important catalytic and regulatory devices, there must be something to be catalysed and regulated. It is the latter system that lies at the core of a primary living system (as opposed to secondary systems, such as multicellular organisms, which are composed of subunits—cells—which are already living). He carefully composed a list of life criteria, and presented a system satisfying them. This system is called the chemoton. One could suspect that this process is circular. Not quite—although the criteria were formulated in a way that helps one to set up a minimal model for a living system, they were nevertheless abstracted from careful examination of contemporary living systems. Again, one could claim that this begs the question: if one looks at *living* systems, then one already has some (maybe tacit) idea of what kinds of systems are alive. Of course, this is true, but it is not problematic. Progress in natural science is an iterative process, and the explanation is ultimately circular as well. If we want to explain all phenomena within the system, then logically we will end up saying that A is explained by B and vice versa. It is the system as a whole that must be coherent and consistent with observation.

4.2   Life criteria: units of evolution and units of life

Thus the first important observation was that, although all contemporary systems operate using the genetic code and enzymes, one can and should do without them for logical (see above) and historical reasons. By the latter we mean that if the origin of life can be solved before that of the genetic code, this makes things much easier. It always pays to solve two difficult problems one at a time, if possible. The second important observation was that one should exclude viruses from the world of the living. This choice seems arbitrary, but Gánti convincingly argues that it is not. He uses an analogy of computation. One can load a program into a computer that orders the latter to produce many copies of the same program. What could such a program do without a computer? Nothing. What could the computer do without the program? It could go on doing any calculation that one feeds into it. In the analogy the living cell is the computer and the program is the virus. It should be intuitively clear from the foregoing that life is in the cell rather than in the virus.

The importance of this step cannot be emphasized enough. It enabled Gánti not to fall into the trap dug by the geneticists. Oparin (1961) defined any system capable of replication and mutation as alive. Most evolutionary biologists would agree with this view. In a sense they are

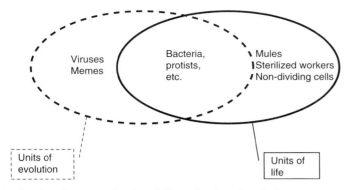

Fig. 4.1   Overlapping sets of units of life and units of evolution.

right; it is systems with these properties that can evolve complex adaptations (purposeful functions) in the natural world, so characteristic of living beings. Many people have accepted the idea that living systems are identical with units of evolution. This conception, treating viruses as living, leads directly to the uncritical acceptance of everything associated with the 'selfish gene' idea. Although Maynard Smith (1986) writes of the dual nature of life (homeostasis/metabolism and heredity), later he reverts to the 'genetic definition' of life. The resolution can be illustrated by the following Venn diagram (Fig. 4.1): whereas all living systems are genetic, not all the genetic entities are alive. This also implies that the applicability of Darwinian principles to the evolution of a population of entities does not necessarily mean that the entities are alive. Just think of Spiegelman's (1970) evolving RNA molecules in a test tube.

Viruses do evolve. In fact they have become one of the most accessible test systems for evolutionary hypotheses (Szathmáry 1992a, 1993a; Turner and Chao 1999). Computer programs can also evolve, as nicely demonstrated by Lenski *et al.* (1999). What, then, is the relationship of units of evolution to units of life? In order to give a tentative answer, one must first define both concepts with sufficient clarity; only after this can one compare the two. Units of evolution must multiply and possess heredity, and the heredity must not be totally accurate (variability). Furthermore, some of the inherited traits must affect the chance of reproduction and/or survival of the units. If all these criteria are met, then evolution by natural selection can take place in a population of such entities (Maynard Smith 1986). Note that this definition does not refer to living systems. Any system satisfying these criteria can evolve in a Darwinian manner.

Units of life as such are less well studied, although cells and organisms are widely known and analysed. Over the years Gánti has refined his 'life criteria' that living systems must meet. He observed, correctly, that reproduction is neither necessary nor sufficient for the individual living state. Many cells and organisms are commonly regarded as alive even if they cannot reproduce (any longer). Potential life criteria (the term 'potentiating' would perhaps be better (Maynard Smith and Szathmáry

1999)) must be met only if the population of units is to be maintained and to evolve. Thus the correct relation between units of evolution and units of life is two partially overlapping sets (Fig. 4.1). I believe that this simple diagram resolves many apparent contradictions in the field. Barrow and Tipler (1988) are right when they point out that potential life criteria (Gánti's term) must be satisfied by living systems if one is looking for the autonomous evolution of a whole biosphere (as an exobiologist, for example).

One can imagine a living system that has been artificially manufactured and could not have arisen by conventional evolutionary mechanisms. Let us assume that it cannot reproduce either. Such a system would still be alive, despite the fact that evolutionarily it is completely detached from the rest of the living world. It is analogous to the 'Garden of Eden' configuration in cellular automata; these are self-maintaining dynamical structures which cannot develop from simpler precursors (Wolfram 1994). If one found such a living being, one could be certain of its unnatural origin.

There is an interesting hierarchy for both units of evolution and units of life; in many cases they do coincide. An organism and its cells are both alive and units of evolution. The latter qualification may be surprising, but it suffices to point out that tumours arise as within-organism selection of genetically moderately unstable cells (Cahill *et al.* 1999). Obviously, such 'selfish' tendencies of the lower level must be suppressed in most cases, otherwise higher-level units would become extinct or would never have arisen in the first place (Maynard Smith and Szathmáry 1995, 1999; Szathmáry and Maynard Smith 1995).

Thus Gánti (1971, 1979a) distinguished between two kinds of life criteria: the absolute and the potential. Whereas the former must be satisfied by any living system, the latter must be shown only by those able to populate a planet. Thus reproduction is a potential, rather than an absolute, life criterion. This elegantly cuts right through the endless debate about whether your grandmother or a mule is alive or not.

4.3   Gánti's concept and some other definitions of life

It is worth drawing attention to a recent investigation by Luisi (1998) of various definitions of life. There is no general agreement about such a definition. Some think that self-replication, enzymatic aid of chemical processes, or cellularity, alone or in some combination, are necessary and sufficient. Clearly, from the logical point of view definitions are arbitrary: one cannot falsify a definition in the same way that one falsifies a hypothesis. So why bother?

In view of this arbitrariness on where to put the marker, is any definition equally good? Surely not, as one definition may be more meaningful than another, depending on what you want to do with it. In fact, the following criteria appear to be important: a definition of life should permit one to discriminate between the living and the nonliving in an operationally simple way and it should not be too restrictive (i.e. the discrimination criterion should be applicable over a large area and should be capable of including life as it is now as well as hypothetical previous forms). All forms of life we know about should be covered by such a definition. Once decided upon, the definition should also help to design experiments on the production of minimal life in the laboratory, consistent with the definition. It should help space explorers in the attribution of the term 'life' to novel biological forms. Finally of course it should be logically self-consistent. (Luisi 1998, p. 617).

The NASA Exobiology Program uses the following working definition of life (Joyce 1994):

Life is a self-sustained chemical system capable of undergoing Darwinian evolution.

The snag is obvious: populations can evolve, individual systems cannot. The definition spectacularly blurs Gánti's important distinction between absolute and potential life criteria. Luisi believes that the following version is even closer to the heart of many, especially adherents of the 'RNA world' idea:

Minimal life is a population of RNA molecules (quasi-species), which is able to self-replicate and to evolve in the process.

This is manifestly worse. The confusion about units and populations thereof remains, and is complemented by another one. Reference to RNA narrows thinking about the problem to such an extent that it becomes intolerable: it elevates an empirical claim about the origin of life on Earth to the status of definition. Suffice it to say that many believe that the RNA world must have been preceded by something chemically simpler (Joyce and Orgel 1999), but then what is the common property of replicating RNAs and those simpler replicators that would qualify both as alive? Not to mention the problem (discussed above) that replicators as such are not necessarily alive either!

Luisi then goes on to consider the following definitions:

Life is a system which is self-sustaining by utilizing external energy/nutrients owing to its internal process of component production

and

Life is a system which is spatially defined by a semipermeable compartment of its own making and which is self-sustaining by transforming external energy/ nutrients by its own process of component production.

The second definition is obviously more restrictive and, as Luisi points out, is consonant with the autopoietic concept (Varela 1979; Varela *et al.* 1981).

It is noteworthy that if one were to build a minimal model satisfying the latter definition, one might arrive at Gánti's self-reproducing spherule, consisting of two autocatalytic subsystems only: the metabolic cycle and the membrane envelope. One could argue, in fact, that this system could be regarded as an 'even more minimal' model for life than the chemoton. One should not forget that it could have some limited genetic properties, resting on possible (hypothesized) alternative forms of the two subsystems (see below). Gánti would respond by pointing out that the limitation on genetics is exactly why we should deny such microspheres the qualification of being alive. Of course, inheritance leads us into the realm of potential criteria, but from the individual's perspective Gánti refers to program control. Thus the complexity of a program which the metabolic or boundary subsystems could store in executable fashion for the operation of the system as a whole is also very limited, if worthy of mentioning at all. That is why it is the chemoton as a whole, rather than some simpler system, which satisfies the life criteria.

4.4   The chemoton satisfies the life criteria

Gánti constructs his model, satisfying both types of criteria, step by step. But why both types? Would it not have been more logical (*sensu* Gánti) to construct a system satisfying only the absolute life criteria? This is an interesting question, and Gánti did not pursue it. His intuitive goal must have been a system that would have been able to populate a planet; otherwise, one could not have used his model for evolutionary purposes. Gánti offers the simplest possible solution to metabolism/homeostasis and control/program in terms of a chemical cycle and a replicating template molecule. The coupling between these two in the basic model is stoichiometric. But the length of the template molecule also has a dynamic effect on the 'cell cycle' of the chemoton. The most primitive way of carrying information by chemical means for the system as a whole is identified with the length of a template molecule rather than its composition, let alone its sequence.

4.5   Compartmentalization and the problem of minimum life

In 1971 nobody had a clear idea of how one could construct a living chemical system. However, everybody assumed that such a formulation must incorporate the genetic code and enzymes. Gánti showed that this assumption, taken for granted by almost everyone (including Eigen), is unwarranted, and then solved the problem that nobody was able to solve. It is no exaggeration to state that Gánti is a visionary. He was a quarter of century ahead of his time. In order to see this, let us look at some developments since 1971. In 1974 (well before Eigen and Maynard Smith) Gánti had already realized that some form of compartmentalization was necessary, because otherwise nothing would hold the subsystems together. The membrane system, the third autocatalytic element, was incorporated into the chemoton, and now the chemoton implies the coupling of the metabolic, genetic, and boundary subsystems.

Recently, the problem of minimum life has been investigated from the bottom up and the top down. In the bottom-up approach researchers like Morowitz (1992) want to design a minimum cell in terms of molecular biology. This is a worthwhile exercise, but logically and evolutionarily comes after, and not instead of, the chemoton, for reasons that must be clear by now. The top-down approach is based on the comparative genomics of bacteria. Researchers are seeking a set of proteins that are common to bacteria with the smallest known genome (Mushegian and Koonin 1996; Hutchison *et al.* 1999). Again, this is interesting, but may be logically flawed for the following reason. Suppose that these genomes consist of four genes: A, B, C, and D in bacterium I, and A′, B′, E, and F in bacterium II. Genes C, D, E, and F show no homology whatsoever. This does not mean that (A, B) and (A′, B′) organisms would be viable. They may have a problem, which must be solved somehow, and it happens to be solved by genes C and D in I and genes E and F II. Thus the chemoton remains the most exciting model of minimal life.

Compartmentalization also plays a vital role in creating a higher-level evolutionary unit. Realistically, a chemoton can utilize dozens of replicating informational molecules. Such molecules are generally expected to compete within the system (Eigen 1971). There must be some other selective force that prevents the systems from dying because of 'intragenomic conflict' (Hurst *et al.* 1996) between the unlinked replicators. Compartmentalization can save the system: chemotons with less distorted, and thus more functional, template composition reproduce faster. Variation in template composition among the chemotons is generated by stochastic effects during replication and chance reassortment of templates into offspring chemotons (Fig. 4.2). Despite harmful internal competition within compartments, group selection among them is a sufficiently strong force to stabilize the population (the 'stochastic corrector model') (Szathmáry and Demeter 1987; Grey *et al.* 1995).

4.6   The autopoietic concept

A second line of thought revolves around the concept of autopoiesis, advocated by Varela (1979). This concept has gained wide popularity in sociology, but has remained unrecognized by many in biology (partly for reasons which also apply to Gánti's work). The approaches of Varela and Gánti are complementary, but the latter is definitely superior in the realm of minimum life and early evolution—it does everything that the former can do, and much more. For one thing, it offers a clear model in terms of chemistry, rendering the problem of 'biological autonomy' tractable.

Related to the concept of autopoiesis is the recent attempt by Fontana and Buss (1994a) to develop an abstract computational and evolutionary chemistry that would model the autonomous evolution of chemical systems. They initially started with lambda calculus, and then went on to try out further calculi. These attempts have met with modest success so far, but they are promising. It is noteworthy in this regard that Gánti realized that in order to deal with the stoichiometry of cyclic chemical

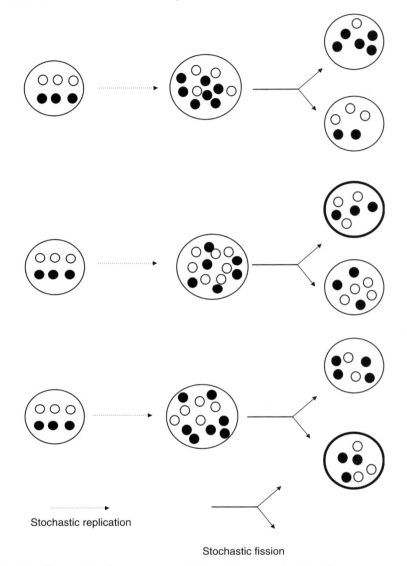

Fig. 4.2 The stochastic corrector model for the group selection of compartments.

systems correctly, one needs a new type of calculus, which he invented and named 'cyclic stoichiometry'. It is also applicable to enzymatic and non-enzymatic networks, and we predict that it will find widespread application in theoretical biochemistry and biotechnology.

A complementary approach entitled 'flux balance analysis' has recently been developed (Schilling and Palsson, 1998). The trick they invented can be explained in terms of the chemoton model. The metabolic cycle produces new A_1, V', and T' molecules. One can analyse its function as if it were not autocatalytic; in this case it would just accumulate these molecules. We are interested in the production of a sufficient quantity of these molecules to construct a new chemoton. Metabolic flux balance analysis uses fairly advanced mathematics and, like Gánti, enquires into what is and what is not allowed in terms of the given stoichiometric constraints of the network. The dynamics of the system (which can be described by chemical kinetics) operates within the stoichiometrically feasible alternatives (there are many such alternatives in the metabolic network of a cell). Flux balance analysis has successfully been applied, for example, to the metabolic capacities of the bacterium *Escherichia coli* (Covert *et al.* 2001).

4.7   The chemoton: a reproducer built from replicators of three types

All subsystems in the chemoton are autocatalytic, and the systems are also autocatalytic as a whole. This 'trinity' of minimal life unites *different* types of replicators. This observation is a resolution of the apparent conflict between Gánti on one side, and Eigen and Dawkins on the other. Maynard Smith and Szathmáry developed and iteratively refined a classification of replicators. The applicability of the most recent version of this (Szathmáry 2000) to the chemoton is as follows. The membrane is a phenotypic replicator (or imitator) because growth and division of the membrane result in a phenotypic relationship between old and new membranes, but the new membrane building blocks are recruited on the basis of their phenotype (the two criteria are an appropriate size and an amphipathic structure with a hydrophobic tail and hydrophilic head). Because replication is based on the phenotype, and because membrane constituents are free to move in two dimensions (i.e. within the membrane), we are dealing with an ensemble type of replicator.

The constituents of the autocatalytic membrane cycle and the genetic template in the chemoton are genotypic replicators, but otherwise they are very different. Autocatalytic cycles of 'small' intermediates, such as the formose cycle (Fig. 4.3) or the reductive citric acid cycle, can have at most limited heredity—all types can be realized in a realistic chemical system. Moreover, their replication is processive rather than modular. One cannot say that replication of an oxaloacetic molecule in the reductive citric acid cycle is 'half complete'. What happens is that after many chemical transformations, the end result turns out to be two molecules of the same kind in place of the one starting molecule. These are holistic replicators; lack of a sequence denies from them the possibility of microevolution. In contrast, the template can be a replicator with unlimited heredity, modular replication, and digital information. It can be a substrate for microevolution, and it is meaningful to speak of the degree of completion of its replication, exactly because it is copied digit by digit (module by module).

Wächtershäuser has argued from 1988 onwards that autocatalytic networks, starting from a very small one, could have evolved before the

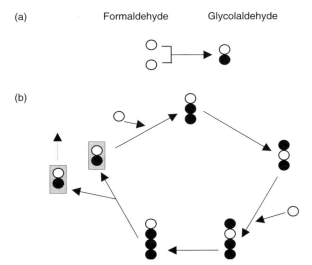

Fig. 4.3 The formose 'reaction', which is in fact a complex network of autocatalytic sugar formation. (a) The 'spontaneous generation' of the autocatalytic seed is a very slow process. (b) the autocatalytic core of the network. Each circle represents a group with one carbon atom.

advent of enzymes and templates. This interesting suggestion—still open to experimental testing—complements chemoton theory and precedes it, both logically and evolutionarily. Remember that we have already stated that being a genetic system and undergoing Darwinian evolution does not necessarily imply that the system is alive. There is no better example that this statement makes sense than here, where we face the possibility of chemical evolution *sensu stricto*.

4.8   Metabolic replicators and the side-reaction problem

There is a severe problem with non-enzymatic metabolic cycles: destructive side-reactions may drain them to such an extent that ultimately the cycle intermediates disappear. This may have been different for cycles on surfaces, but we do not know (yet). As King (1982, 1986) has pointed out, the smaller the cycle, the better are the chances for its propagation. Suppose that there is a simple autocatalytic cycle of n steps (similar to the system in Fig. 4.3, where $n = 5$). At each step one has the legitimate reaction leading to the next cycle intermediate, and a number of side-reactions that drain the system. The latter give rise to all sorts of side-reactions. Let the specificity of a reaction at step i be s_i, which is the rate of legitimate reaction divided by the total rate of all (legitimate plus side) reactions. Successful growth of the cycle is guaranteed if

$$2\prod_{i=1}^{n} s_i > 1 \qquad (4.1)$$

or, if we calculate using the geometric mean σ of the specificities,

$$\sigma^n > 0.5, \qquad \text{i.e. } n < -\log(2)/\log(\sigma). \qquad (4.2)$$

This shows that the viable system size n increases hyperbolically with specificity.

A result obtained for somewhat analogous systems (reflexively autocatalytic protein networks (Kauffman 1986)) shows that the larger the system, the more likely it is to obtain cyclic closure and autocatalysis for the network as a whole. However, we have just seen that as system size increases, the chance of side-reactions also increases. Nobody has yet provided a satisfactory solution to this 'paradox of specificity' (Orgel 2000; Szathmáry 2000).

Wächtershäuser's idea of surface metabolism may save the case, because of the benign catalytic effects of an appropriate mineral surface. Orgel (2000) forcefully questions this possibility, essentially because a fairly large network consists of reactions of different types, which would require a multifunctional mineral surface akin to a multi-enzyme complex. Whether such mineral surfaces can exist is an open question. Alternatively, the primordial autocatalytic cycle must have been very simple and small.

This means that the nonenzymatic chemoton may be unfeasible. However, this does not invalidate it as a model of the *regulated* system. What is more, a more developed version of the chemoton assumes that members of its template population exert catalytic activity on the metabolic cycle. In the terrestrial context, this could have been the era of catalytic RNA.

4.9   Enzymatic RNAs (ribozymes)

How can the gap between the non-enzymatic basic chemoton model and prokaryotes with protein enzymes and a genetic code be bridged? Gánti, again ahead of his time, took suggestions by Woese (1967), Crick (1968), Orgel (1968), and White (1976) seriously, and devised the catalytic chemoton model in 1979. He argued that a 'missing link' could be provided by a system in which the reactions of the metabolic cycle were catalysed by enzymatic RNA molecules belonging to the now varied pool of sequences in the genetic subsystem. He even coined the term eRNA ('enzyme RNA', cf. tRNA, mRNA, etc.). Thus he assumed that RNA molecules could have a general catalytic capacity and could have functioned in primordial organisms. He also pointed out that there would have been strong selection for such function in the chemoton. Gilbert's description of an 'RNA world' appeared in 1986, and the metabolically complex RNA world of Benner and his colleagues was described in 1987 and 1989. We can trace back the metabolically complex RNA world idea to White (1976), who argued that nucleic acid co-enzymes could be fossils of an earlier metabolic stage, and Gánti (1979), who demonstrated how replicative ribozymes (the term now used for his eRNAs) could have been selected for in chemotons.

It is reasonable to ask how far the logical build-up of the chemoton, as outlined in the present book, could aid one in the understanding of the *historical* process of the origin of life. Gánti took an essentially Empedoclean standpoint: he imagined potential subsystems of the chemoton to have arisen independently, followed by selection of the various combinations. This view may be mistaken. Although there have been promising attempts to synthesize chemoton-like systems (such as the production of self-reproducing micelles with internal template replication by Luisi *et al.* (1994)), the successful incorporation of a non-enzymatic metabolism may turn out to be impossible. In contrast, several workers in the field share the view that precellular evolution could have gone as far as the appearance of ribozymes, before the spontaneous generation of chemoton-like systems (Maynard Smith and Szathmáry 1995). If this view turns out to be correct, the most primitive chemotons were ribozymic right from the beginning.

4.10   Protocells without a metabolic subsystem: the ultimate heterotrophs?

Szostak *et al.* (2001) have outlined a research programme aimed at the synthesis of life in the form of a simple cell. The proposed system consists of an autocatalytically growing and dividing membrane, and two specific ribozymes. The first ribozyme is a general replicase, able to replicate other copies of itself as well as the second type. The second ribozyme takes part in lipid synthesis. Thus the coupling between templates and membrane is catalytic. Complex raw materials, such as the monomers for RNA replication and lipid precursors, are taken up from the medium. Some experiments on reproducing vesicles (Bachmann *et al.* 1992; Luisi *et al.* 1994; Walde *et al.* 1994) are obviously moving in this direction.

It may be possible to build such a functioning system, although membrane transport and synchrony between template replication and membrane growth pose special problems. (Remember that in the chemoton the basic coupling between these two subsystems is stoichiometric.) Could it be considered alive? We leave this quesiton as an exercise for the reader.

5 The philosophical significance of Gánti's work

James R. Griesemer

5.1 Units and levels: biology and philosophy

What are the basic kinds of entities to which the laws and principles of biology apply? Two traditions—structuralism and functionalism—have emerged over the last 30 years to answer this question for evolutionary theory, which plays a fundamental interpretive role in the biological sciences. Both traditions take for granted the chemical basis of contemporary living things, a practice that limits what evolutionary theory can say about the origin of life and about evolutionary transitions producing new levels of organization. The theoretical biology of Tibor Gánti, developed over this same period, offers new possibilities for answers to the questions about units and thus can ground a fresh approach to the philosophy of biology.

The units of evolution problem resembles other units problems in science. Identification and classification of the subjects of research are necessary steps in theory-building and are also important when assessing whether theoretical descriptions in the science are well formed and accurate. Theories cannot be built or checked unless they have identifiable subjects. Quantitative theories need precisely specified quantitative units so that equations 'balance'; for example, if the left-hand side of an equation is in units of distance divided by time, then the right-hand side must be as well.

The philosophical challenge in evolutionary theory over the last 30 years has been to convince biologists they even *have* a units problem. Since the evolutionary synthesis and the molecular biology revolution of the mid-twentieth century, it has seemed clear enough that genes mutate, organisms are selected, and populations evolve, except in odd and esoteric cases such as when epigenes mutate, groups are selected, or ecological communities evolve. The argument that our general outlook on biological theory must change as a consequence of getting these sorts of units questions right has made some inroads into biological consciousness. However, to most researchers, units analysis is dull philosophical spade work that is pedestrian and unilluminating except, of course, during times of dramatic failure of principle or application of the sort that leads to scientific revolution (Kuhn 1970). Two typical reactions to the claim that evolution applies to units at multiple levels of organization are as follows: (i) since evolution is change in gene frequency, it is unnecessary to represent effects due to higher levels in order to make correct measurements of evolutionary change; (ii) even if representation of effects at higher levels improves the causal picture of evolutionary change, the effects are rare or of such small magnitude that they can be safely ignored in calculations. (See Sober and Wilson (1998) for a review of these attitudes and Wimsatt (1980) for a general discussion of heuristics in the units of selection controversy.) The argument of those scientists who do incorporate serious attention to units analysis into their work is that these claims require investigation—serious engineering analysis—before they can be adopted as heuristics. We need to know when and where they break down in order to endorse them. But to do that we must incorporate them into the models and study them in a broad analysis of units.

It is standard engineering practice to understand how something works by studying what makes it fail. To detect failure, a study of what the engineer's blueprint specifies as success is first required. In empirical science, blueprint specifications of how nature ought to work come in a variety of forms, including theoretical models, experimental designs, and observational protocols. An exact theoretical science must involve engineering analysis of theoretical models because these are the blueprints that lead to theoretical predictions as well as explanations of what the research process discovers. Units analysis makes a good problem for philosophers of science because failures at this level point to the kind of fundamental issues that philosophers seek to address. Thus their study may reveal the nature and limits of scientific concepts, principles, laws, methods, and practices. Hence units analysis can serve investigations into the power and generality of theoretical principles to predict and explain.

The units of evolution is a key problem in the philosophy of biology. Not only is evolution an important field of biological research, but evolutionary units are of general significance because evolutionary theory is central to the whole structure of biological theory. Evolution integrates biology and provides its subjects. As Dobzhansky wrote, 'in biology nothing makes sense except in the light of evolution' (Dobzhansky 1970, 5–6). Conversely, much of the rest of biology must figure in any significant understanding of evolution. Thus a proper account of the units of evolution provides a key not only to evolution but also to the conceptual analysis of biology as a whole.

Philosophers are interested in the units of evolution in part because examination of units, their place in theory structure, and their role in empirical successes and failures all provide windows on the nature and practice of science (Wimsatt 1980). Theoretical and empirical biologists are interested in units of evolution for the same reasons. If units analysis fails, something might be wrong with theory. This was the thrust, for example, of early work on group-level selection: models describing evolution at gamete and organism levels failed to explain changes in gene frequency, whereas models including the group level fitted the data better (Lewontin and Dunn 1966). A parallel case about units of heredity has been made in the light of new evidence of epigenetic inheritance systems (reviewed by Jablonka and Lamb (1995); see also Griesemer (1998)). If something is wrong with theory, the units analysis checkpoint may indicate steps toward theoretical advance. Gánti's work investigates life criteria for the units of an exact theoretical biology, and thus is of high significance for philosophical investigations. 'Exact' here means 'admitting of absolute precision', which is only possible in the abstract world of a model—balls can move on billiard tables without friction in theoretical physics models, organism mating can be perfectly random in theoretical genetics models—and it is theoretical science which produces and investigates models.

Of the two main traditions of units analysis in evolutionary theory, the structuralist tradition identifies evolutionary units with things appearing at particular levels in the hierarchy of biological organization (Lewontin 1970; Wimsatt 1980; Hull 1988; Lloyd 1988; Brandon 1990). This hierarchy is bracketed below by the entities of physics and chemistry (atoms and molecules) and above by the entities of the social sciences (organized social, cultural, and linguistic communities). The biological entities in between include biochemicals (e.g. genes and enzymes), supermolecular complexes (e.g. organelles), cells, tissues, organs, organisms, and some kinds of groups of organisms (e.g. families, demes, species, and ecological communities). The evolutionary units question in the structuralist tradition is: Which of these given kinds of organized entities, under what conditions, are subject to forces of evolution or natural selection? These forces are specified by explicit evolutionary models which describe the dynamics of ideal units in populations.

The functionalist tradition argues that things are not units of evolution by virtue of their specific kinds of matter or location in the hierarchy of structural organization, but in terms of the functional role they play in the evolutionary process, regardless of the level of organization at which they occur (Dawkins 1976, 1982; Hull 1981, 1988). The functionalist alternative to the structuralist tradition emerged in the 1970s when units of evolution were interpreted directly in terms of two basic functions, replication and selective interaction, as a way of comprehending (or challenging) the extention of evolutionary theory to entities at many levels of structural organization besides that of organisms—the gene below and the group above (Sober 1984). These concepts generalized the functions of genotypes and phenotypes in classical genetic models of evolution.

Functionalists agree that function rather than structure makes something a unit of evolution (or selection or heredity) even though they disagree about how to characterize the functions and what are the empirically significant functional units in evolution. For example, Dawkins says that 'replicators' are the significant units of selection, while Hull, Brandon, and others say that 'interactors' are the significant units. Entities at each level of the structural hierarchy can be characterized in terms of both kinds of functional role in evolution (and in physiological, developmental, and ecological processes as well). The empirical question for the functionalist is: When and under what conditions do entities at a given structural level fulfil the roles of replicators and interactors?

Something plays the evolutionary role of a 'gene' if it has the properties of a replicator: longevity, fecundity, and copy fidelity (Dawkins 1976). A thing does not necessarily play that role because it has the structure of a molecular chain of nucleic acids. Indeed, on functional grounds, many chains of nucleic acids are not genes (e.g. the sequences used to prime the polymerase chain reaction in PCR machines). Whether entities at other levels of the hierarchy can play the role of genes in evolution—for example, germinal (or even somatic) cells, cell lines, tissues, organisms, demes, species, or ideas—is an open empirical question quite different from the conceptual problem of analysing the nature of replicator function.

Structure and function perspectives need not compete, although the relationship between them has not been fully examined (Griesemer 2000b). Specifically, replicator and interactor functions are not really defined independently of the structural hierarchy, insofar as we discover and analyse the properties of these two functions by investigating the structures of modern living things which happen to fulfil those functions today. Genes are the paradigm replicators from which we learn what replicators must be like. Organisms are the paradigm interactors from which we learn what interactors must be like. Replicators must indeed be gene-like because genes *are* replicators, but it does not follow from this that the properties which geneticists use to describe modern Mendelian factors or highly evolved DNA sequences are *necessary* properties of replicators. Interactors must be organism-like because organisms are interactors, but it does not follow that the properties of the kinds of organisms that biologists have studied are necessary properties of inter-actors (e.g. the fact that they have physical boundaries or integuments such as skin or bark). Hull (1988) pointed out that there is a substantial 'vertebrate bias' in the types of features that biologists, who are all vertebrates, take to be 'typical' of organisms, even though vertebrates are very 'atypical' organisms (see also Buss 1983, 1987). Modern genes may be just as atypical of replicators in general as vertebrates are of organisms in general; it is an empirical question with implications for units analysis. If there are replicators besides the ones made of DNA, then genes are not *the* units of heredity. The fact that all *modern* organisms contain nucleic acid replicators (DNA or RNA) is no assurance that replicators must be made of nucleic acid, that most are, or even that they are required for life. Gánti's focus on the role of 'genetic molecules' in the regulation and control of an autocatalytic metabolism makes it clear that replicator function need not involve all the properties of modern DNA.

Replicator functioning in the modern biological world might turn out empirically to be fulfilled only or mainly by genes and in the cultural world by 'memes', as Dawkins argued. But evolution *might* have proceeded from different origins and to a different result. A proper theory of evolutionary units would distinguish such possible but supposedly non-actual courses of evolution from conceivable but impossible ones. What is possible *today* might not have been possible at the origin of life, and what was actual at the origin might not be possible today. A definition of the replicator function must cover both situations. If the replicator function is defined in terms of properties of DNA molecules (contemporary or ancient), then kinds of replicators unlike DNA would be ruled out by definition, even though they might have existed at the origin of life before the evolution of DNA replicators.

Unit concepts of the replicator are like the proverbial leaky boat that must be rebuilt at sea. Each new wave of empirical evidence is likely to upset the delicate balance of the barely floating boatload of definitions. But the boat must float at all times if its passengers are to survive; its current condition is a severe constraint on what the reconstruction of a unit concept can be like. We use current conceptions of genes to tinker with the idea of the general replicator function in evolution. Since gene concepts have mostly been built around the experimental study of genes in living organisms, the conceptual resources for thinking about replicators have been shaped by the genetic boat floating in modern evolutionary waters. However, when the evolutionary *origin* of genes and genetic systems is addressed, the limitations of traditional approaches to the philosophy of evolutionary units becomes apparent. *One of Tibor Gánti's most significant contributions to theoretical biology is to provide a viable means of thinking about living units without prior commitment to modern genes. Gánti's chemical perspective on life provides the continuity that keeps the boat floating.*

The discovery in the 1980s that RNA can have catalytic properties was quite startling, given prevailing views about the genetic role of nucleic acids and the catalytic role of proteins. At the time, many thought that nucleic acids could not be enzymes because enzymes were proteins *by definition*. Might it be that attention to modern DNA has similarly distracted us from the general properties required for replicators? Are epigenetic inheritance systems replicators? Could intermediary metabolism be a replicator? Is a cell membrane a replicator? Are whole developmental systems replicators? These are certainly not 'digital' or modular systems of unlimited heredity in the strict sense that DNA replicators are (Szathmáry 2000), but it is not clear whether the digital property of unlimited heredity is necessary for replicators to be evolutionarily significant or only a contingently irreversible property that modern DNA replicators have evolved from non-modular holistic replicator ancestors. How should the replicator concept be formulated in order to serve a completely general analysis of units of evolution?

Defining the replicator function in terms of properties of (modern) DNA would not be a serious limitation on the functionalist approach to units of evolution if the theory were only applied and checked within the framework of a fixed structural hierarchy of contemporary biological organization. Differently put, as long as evolutionary theory concerns the functioning of *contemporary* units at *fixed* levels of the biological hierarchy, i.e. the subject of most evolutionary and other biological research, the functionalist approach may be adequate to its intended task. However, if a philosophy of units is to address problems going beyond this scope—for example to problems of evolutionary *transition*, i.e. of the evolutionary processes from which the contemporary hierarchy of life originated, including the origin of life from non-life—then a different approach is needed (Maynard Smith and Szathmáry 1995). Otherwise, we risk assuming what is to be proved. This threat of circularity is the philosophical risk that conceptual projects like Gánti's must overcome. On the one hand, they must offer criteria that delimit units (e.g. of replicators or of living things), but at the same time they must not do so in a way that prevents their rejection if they turn out to be conceptually or empirically inadequate.

If we assume a fixed structural hierarchy or define replicators in terms of contemporary structures (nucleic acids) in order to specify units, these problems cannot be solved. Theories of origins must not assume, in their concepts and principles, the existence of the very things and properties whose origins are to be explained. The process of evolutionary transition *creates* levels of organization, and so these levels cannot be assumed by a theory of evolutionary transition. If nucleic acid replicators are highly evolved (as Eigen's paradox and hypercycle solution shows they must be and which Gánti's theory addresses), then there must have been other less evolved kinds of replicators than contemporary DNA-based genes. But if there were, the replicator concept should not appeal to properties of contemporary genes without establishing that they are *also* properties of pre-transition proto-genes as well unless the scope of the replicator concept is not all life, but only all modern life.

The problems of origin of life and evolutionary transition highlight the weaknesses of traditional approaches to the units of evolution; thus work on these 'esoteric' phenomena is theoretically and philosophically significant despite the temporal remoteness of the origin of life and the low relative frequency (i.e. zero) of life's original systems among contemporary organisms. If theory is to be general, then it must cover remote times as well as esoteric levels of organization. The common dismissal of a phenomenon as infrequent and therefore irrelevant to the formulation of 'good enough' theory no more applies in biology than it does in physics. Where would physics be if it ignored the 'big bang' just because it concerns only a few early moments in the history of the Universe?

Structuralists cannot explain the evolution of the hierarchy because they assume its existence when they define units. Functionalists cannot explain the origin of life from non-life because they assume the presence of the life functions (replication and interaction) when they define units. A solution to such problems of biological origins requires understanding relationships between contemporary biological units and systems of non- and proto-biological units from which contemporary biological units evolved. In short, current perspectives in philosophy of biology are too limited for the job because they are shaped by an understanding of only the contemporary highly evolved world.

A philosophy of biological units suited to esoteric as well as typical problems of evolutionary biology is greatly facilitated by the perspective of Gánti's chemoton model and life criteria (Gánti 1979b, 1987, 1997) (reviewed by Maynard Smith and Szathmáry 1995, 1999). This perspective adds chemistry to the traditional biological and evolutionary units approaches and offers a method for analysing the relationship between chemical and biological units. Gánti offers several key ideas that should figure prominently in a renewed philosophy of evolutionary units: fluid state automata, cycle stoichiometry, the chemoton model, and an integrated approach to life criteria and models. These are discussed in many of his publications, including *The Principle of Life*, only a few of which have been available in English translation or English language scientific periodicals. Their significance for the philosophy of biology will be considered in the rest of this chapter.

5.2   Self-reproducing automata in the fluid state

In the 1950s, von Neumann offered one of the first theoretical investigations of self-reproduction (von Neumann and Burks 1966). Gánti describes self-reproducing *fluid* state automata that are quite different from von Neumann's solid state automata. von Neumann imagined self-reproduction in the physical terms of solid state electrical machines which carry a description of the structure of the machine including a constructor capable of duplicating both the description and the constructor. von Neumann's approach is ingenious, but Gánti argues that solid state machines could not physically satisfy conditions for self-reproduction while fluid state machines could. Gánti's chemoton is a model fluid state automaton capable of self-reproduction. Moreover, there is no complete description of a chemoton contained within it. The reproduction of the chemoton is an emergent property of the chemical behaviour of the system.

To see why fluid machinery is required for the living state and self-reproduction, a chemist's perspective is very helpful. Self-reproduction involves the production of a new offspring entity with relevantly similar properties *out of* the old parent entity. These properties must include the capacity for self-reproduction in order for the production of the offspring to be a *re*production (Griesemer 2000a,b,c). The flow of organized matter conferring developmental/constructive capacities from parent to offspring, in contrast to mere repetition of pattern, turns out to be a very hard problem to solve in the solid state because the corresponding parts of parent and offspring hinder one another geometrically while the latter is being produced. Even inorganic crystals can *transmit* pattern by interaction with adjacent monomers, and so self-reproduction must be more complex than crystal growth (Penrose 1959; Szathmáry and Maynard Smith 1993).

Self-reproducing entities must be complex, but to be complex they must also be fairly large. Being large and also solid state means that the production of the offspring would have to occur adjacent to the parent rather than within its body to avoid geometrical hindrance. These conflicting constraints on size and complexity due to the rigid geometry of the solid state are not present in the fluid state. Different 'structures' can be mixed together and share the same geometrical space, and yet be chemically and dynamically distinct, just as two populations of birds of different species can flock together in the same airspace and yet remain distinct populations. Offspring and parent chemical cycles can coexist geometrically for a time without disrupting one another. In Gánti's chemoton, a metabolic cycle can double in size and divide, with the two prospective offspring systems residing inside the same membrane without getting in one another's way. In modern cells, the 'dance of the chromosomes', in which the parental chromosomal material duplicates and moves to opposite sides of the cell, takes place within the body of the parent cell. The duplication of the chromosomal material does not disrupt the basic functioning of the cell as a whole. The architectural constraints on the solid state are too severe for von Neumann's automata to be a plausible abstract model for life. It is no accident that cells are basically fluid systems, even though they include many solid state component structures.

In the decades following von Neumann's work and in parallel with that of Gánti, a new perspective on biological self-organization was developed by Maturana, Varela, and colleagues around the concept of 'autopoiesis'. Auto-poiesis, self-production, is described by a theory of biological organization, i.e. a theory of relations among component parts that define the system comprising them as a unity. A unity is 'an entity (concrete or conceptual) separated from a background by a concrete or (*sic*) conceptual operation of distinction' (Maturana 1981, p. 15). The autopoietic tradition developed by Maturana and Varela focuses on this dynamic characterization of organizations, in contrast with the static aspect of structure or geometrical relations among components comprising a whole. (Another more recent program investigating dynamic organizations, focusing on the emergence of autonomy, has been developed by Hooker and colleagues (e.g. Collier and Hooker 1999). Simulation studies of the emergence of organization based on theoretical models of abstract chemistry are explored by Fontana and Buss (1994a,b) and Fontana *et al.* (1994).)

Gánti also focuses on the dynamic aspect of life in his chemoton theory, emphasizing that living organization is best understood in terms of *fluid* machineries whose 'inherent unity' is not constrained by the geometry of solid state structures. Inherent units, for Gánti, are systems whose properties are not additive compositions of the properties of the parts, but rather whose new qualitative properties are 'determined by interactions occurring according to the organization of the elements of the system' (Gánti 1987, p. 69) (see p. 76 of this volume). Wimsatt (1985) has given an analysis of emergent system properties in terms of various kinds of failures of aggregativity.

The distinction between organization and structure is critical to the insight shared by these traditions because self-organization, self-maintenance, and self-reproduction cannot take place within the structural space of the entity in the solid state. von Neumann's self-reproducing automata, for example, required a 'constructor' that was part of his hypothetical machine but geometrically outside the space of the part which carried the information for the machine's structure (von Neumann and Burks 1966). In the end, von Neumann's automata are only descriptive models; they cannot be physically realized because the geometrical constraints of the solid state prevent the reproduction of the machine in its own space. Thus, not only is reproduction not an absolute requirement for living systems (see also Varela *et al.* 1981, p. 8), but understanding reproduction depends on an account of these requirements, which Gánti calls 'real' life criteria for living systems. Both the autopoietic tradition and the chemoton theory emphasize this.

Another key similarity between autopoietic and chemoton theories is their recursive structure. Varela and colleagues state:

The autopoietic organization is defined as a unity by a network of productions of components which (i) participate recursively in the same network of productions of components which produced these components, and (ii) realize the network of productions as a unity in the space in which the components exist. (Varela *et al.* 1981, pp. 7–8).

They go on to identify *autonomy* as the product of the operation of autopoietic organization, thus the realization of autopoietic organization is *self*-realization.

Gánti likewise discovered a recursive structure of self-realization in the autocatalytic chemical cycles of chemoton organization. The organization of such cycles is a unity in the autopoietic sense because it is due to a network of chemical productions of cycle components which participate recursively in the same networks that produced them, and these are realized in the space, bounded by a membrane, in which the components exist. However, Gánti's theory goes beyond autopoietic organization in his stoichiometric coupling of an autopoietic metabolism to the bounding membrane and a genetic molecule. Stoichiometry introduces *flow* of matter as a fundamental property of biological systems, one which can be analysed from a more realistic chemical perspective than the cellular automaton models of Varela and his colleagues. In this way, an operating chemoton also realizes self-reproduction because its autopoietic operation leads to the formation of new autopoietic entities. While both Gánti and the autopoietic tradition argue, correctly and insightfully, that reproduction and evolution are not requirements of living systems, Gánti's chemoton theory reveals a deep connection between the absolute or real life criteria—those corresponding to the conditions for autopoietic organization—and the potential life criteria of growth/reproduction, heredity/evolution, and mortality in the chemical processes leading to duplication of chemical cycle components.

By virtue of this connection, Gánti's work provides a conceptual bridge between the analysis of autopoietic organization and the autonomy of living systems on the one hand, and the analysis of units of evolution on the other. His chemical perspective on biological organization focuses attention on material organization and stoichiometric relations as the basis for identifying and individuating biological systems.

Progress in the philosophical investigation of units of evolution would be considerably enhanced by incorporating chemoton theory because it is precisely the problem of what counts as a relevantly biological system that has plagued the rigorous analysis of evolutionary units. Taking the properties of living systems for granted in the service of an account of units of evolution necessarily begs the important question of the evolutionary origin of those units. Moreover, analysis of units of reproduction, building on Gánti's insight into the chemical conditions of biological organization, has revealed the deep connection between heredity and development, and so the integration of work on units of evolution with developmental systems theory could also be advanced (Griesemer 2000a,b,c) (see also Szathmáry and Maynard Smith 1997). The cutting edge in philosophy of biology today concerns the integration of evolution with development, and Gánti's chemical perspective is a much needed complement to these process-oriented but purely biological perspectives.

5.3   Cycle stoichiometry: the language of living systems

It is characteristic of life that it operates in cycles, and so taking a chemical perspective on life calls for an investigation of the chemical basis of cyclic behaviour of living things. Gánti developed cycle stoichiometry in order to describe the kinetics and dynamics of chemical autocatalysis and self-reproduction. It is a perspicuous notation for representing reaction cycles in intermediary metabolism, membrane neogenesis, and information–macromolecular replication. In this notation abstract models of the cell can be described as super-systems of interconnected auto-catalytic chemical cycles, bringing basic biological units into the chemical perspective. As a notation, cycle stoichiometry has the great advantage of distilling and focusing on aspects of chemistry that are particularly important for a broad understanding of biological subsystems, super-systems, and units. Thinking in its terms can also serve as a heuristic tool for extracting illustrative principles from chemical reaction networks, while providing a rigorous methodology that is capable of handling the complexity of real biochemistry.

Cycle stoichiometry is discussed in greater detail in Gánti's technical work than could be presented in this book (Gánti 1984, 1989, 1991). More importantly for philosophy of biology than these technical details, however, cycle stoichiometry provides language appropriate for describing, in functional terms, the kinds and patterns of chemical reactions that can satisfy life criteria in abstraction from many of the chemical details. Cycle stoichiometry makes reductionism look philosophically promising again as a heuristic research strategy (Wimsatt 1997; Griesemer 2000a) to complement, rather than compete with, holistic approaches such as developmental systems theory (Griffiths and Gray 1994; Oyama and Lewontin 2000; Oyama *et al.* 2001). A key point of philosophical signifi-cance of Gánti's book is its devotion to showing that cycle stoichiometry describes life's functioning at its chemical minimum and therefore at its most general biologically.

5.4 The chemoton

Gánti's model of the simplest chemical supersystem capable of self-reproduction is called the chemoton. The chemoton, Gánti argues, satisfies five absolute life criteria, i.e. criteria that must be satisfied at each moment a thing is in the living state—inherent unity, metabolism, inherent stability, information–carrying, and program control. Gánti identified three 'potential' life criteria that at least some things in the living state must also satisfy if there is to be an evolving living world—growth/ reproduction, capacity for hereditary change and evolution, and mortality. The potential life criteria are related to requirements for units of evolution.

Szathmáry considers Gánti's distinction of absolute life criteria for the living state from potential life criteria for units of evolution to be an important insight. Gánti argues that satisfaction of the potential life criteria is not necessary for satisfaction of the absolute life criteria, and so cases like sterile organisms (which are alive but cannot reproduce) and aestivating organisms (which are in 'suspended animation' and therefore neither alive nor dead) can be handled. I agree, although I also think that it might turn out that satisfaction of either set of criteria by members of a population of entities may be *sufficient* for satisfaction of the other. This is a conceptual problem that should interest philosophers.

The chemoton model identifies three basic autocatalytic component subsystems which are present in all living cells: a metabolism, an information-carrying subsystem, and a membrane. In the simplest form of the model presented in *The Principle of Life*, Gánti showed how these three can be combined stoichiometrically to form a self-reproducing chemical autocatalytic cycle. Cycle stoichiometry makes this demonstration efficient, understandable, and heuristic for a wide variety of problems about units of life. The metabolic subsystem of the chemoton is similar in important respects to known autocatalytic chemical cycles, which indicates that the model holds promise for interpreting the origin of life. Because it satisfies Gánti's absolute life criteria, the chemoton model can also be used heuristically in the philosophy of evolutionary units.

Life criteria can guide the search for such units. However, there are two ways to search: (i) look up and down the structural hierarchy for living units that satisfy functional or structural criteria for evolutionary units, or (ii) design a minimal unit out of units that are themselves non-living (i.e. a chemical unit), a population of which satisfy the potential life criteria. The first approach is descriptive while the second approach is constructive. The first approach is basically a matter of heuristic abstraction and iteration from what we already understand as life: study obvious cases of living things (such as organisms like us); formulate criteria based on these easy cases; test and refine the criteria on more troublesome cases (like bacteria, viruses, schools of fish, the Earth's biota) to make them easier cases; repeat. The second approach is also iterative, but it does not simply abstract from what we know and believe about the easy cases. It does not take for granted that we always know life when we see it. Rather, it constructs a model on the basis of what we know and believe we can do with simple concepts and principles. Life criteria are first formulated for the easy cases—what we think they are like—but then it seeks to construct something, a model, that actually satisfies the criteria. If it can be built, so the engineering ethos goes, then the techniques of construction can be applied to harder cases.

The units of evolution controversy in philosophy of biology has tended to limit thinking about the functioning of evolutionary units within certain channels. Its effect on our understanding of the logic of evolution is similar to the teleological views of evolutionary adaptation that have been so heavily criticized by philosophers. We can no more assume the organizational hierarchy in general evolutionary theory than we can the end-state, or 'telos', for the sake of which natural selection acts. Natural selection is not a forward-looking teleological process, nor is it an 'upward'-looking process. Complexification and evolutionary transition to new levels of organization are to be explained, not assumed. *Defining* hereditary units in terms of what we know about *contemporary* genes, for example, is teleological in this sense.

Teleological assumptions tend to creep into the functional analysis of biological units as well because functionalism (in evolutionary units analysis) shares many assumptions with the structuralism it was introduced to correct. The belief that only DNA has the structure necessary to function as (biological) genes, if based on a definition of the gene in terms of the structural hierarchy, risks explanation of lower levels of functioning in terms of higher levels of structure which have not yet evolved. Note that after higher levels of structure have evolved, there may even be effects of higher-level evolution on lower levels of structure, further limiting the value of studies of contemporary lower-level units as guides to general properties of primordial replicators. The hindsight of teleology limits scientific imagination to such an extent that all the theoretician can do is sail along familiar shores, while what is needed is an exploration that strikes out across the ocean of uncertainty about processes of evolutionary transition to new levels of organization.

5.5 Lessons from Gánti's work: engineering models and life criteria

Usually, life criteria characterize units by describing them. Descriptions can be useful for interpreting difficult cases (cellular slime moulds, dormant spores, mules) or borderline cases (viruses). However, they tend to be limited by those things that we already recognize as living and the terms in which we recognize life (DNA-containing etc.). Therefore life criteria which merely describe some living things are of limited utility. Exceptions can be found to most simple 'life functional analyses' of life, and so life criteria do not serve very well as conceptual analyses or definitions, but at most as criteria for practical decision-making (Feldman 1992). Many of the weaknesses of traditional approaches to life criteria can be traced to their development by purely descriptive means. Cycling between descriptions and tests of descriptions resulting from further observation and intuition is not a very robust procedure.

An engineering or modelling approach is to develop life criteria as part of an iterative process working between description and construction. Life criteria which describe (some) living things can be used to construct theoretical or working (chemical) models of living things by virtue of satisfying life criteria to a first approximation. Then the behaviour of these models can be used in turn to refine the criteria in terms of how and how far the model's behaviour diverges from that of (known/suspected) living things. To the extent that the materials and methods of construction depend on different, deeper, or emergent properties of life than those available for convenient description, construction contributes positively to the project of constructing life criteria.

Gánti pioneered the constructive use of life criteria as part of a programme for the development of an exact theoretical biology. What makes his approach appealing is twofold. First, the chemical perspective leads to a simple heuristic model, the chemoton, with which to think through general descriptions of life criteria. Second, emergent properties of chemoton behaviour discovered by computer simulations are found to be characteristic of life as well. The chemoton goes through a life *cycle* as a function of the dynamic behaviour of its stoichiometrically coupled subsystems. The growth, development, and reproduction of the chemoton as a whole is emergent behaviour, which is not built into the model blueprint.

Gánti's real life criteria are similar in some respects to 'life functional' approaches tracing back to Aristotle (Feldman 1992). If Gánti had finished his 1971 book at Part II, Chapter 5, 'The criteria of life', we would have a clever philosophical scheme for 'defining life', but all we could do at that point would be to compare the criteria against further cases that might be observed.

The rest of Gánti's book, from Chapter 6 onward, is quite different in character from the conceptual work of a philosopher, although it does not neglect the nuances and fine points of the life criteria offered in the early chapters. Rather than simply look and see whether the criteria are satisfied by more examples of biological things, Gánti describes a hypothetical construct or model of a chemically minimal system, the chemoton, which satisfies the life criteria. Gánti's important innovation in 1971, contemporary with Eigen, was to see how to formulate a minimal model of a living system constructively. What this means is that the model is based on known chemical principles in such a way that, at least in principle, a real system could actually be synthesized in a chemical laboratory with all the properties of a chemoton from constituents that do not satisfy the life criteria. In fact, chemists are attempting precisely this (von Kiedrowski 1986; Rebek 1994; Szathmáry 2000). The achievement of a synthetic chemical biology would be as great as that of organic chemistry a century ago.

It might be thought that this constructive strategy and achievement has already been reached. There is a large experimental literature on the origin of life (Deamer and Fleischaker 1994). Since Urey and Miller in 1953, many experimentalists have investigated chemical reactions that suggest at least the outlines of a synthetic evolutionary biochemistry. However, synthesis of (some of) the constituents of life is not yet (or not even) an attempt to construct a system that satisfies life criteria. What Gánti realized is that part of the problem of constructing an appropriate chemical model of life is one of a chemical *notation* appropriate to biology. Cycle stoichiometry fills that gap by providing a formal language in which to link chemical models to life criteria.

The chemoton theory offers a number of lessons for the philosopher of biology interested in thinking about life in its most general abstract aspects. Because the heart of chemoton theory is an abstract chemical model of living organization, these lessons should be understood more as tools for disciplining theoretical investigation than as summary truths about empirical nature. Two points will be discussed.

5.5.1   Control is distributed

In modern cells, the metaphor that DNA is the master molecule of the cell describes the common biological dogma that genes are the central controllers of cell function through their role in 'coding for' protein sequences (Sapp 1987). The language of 'control' is used primarily in popular literature rather than in scientific publications, where 'switch' or 'selector' gene is favoured (Robert 2001, p. 289). However, that dogma should be treated as something to be explained, a product of early stages in the evolution of life, and not as an assumption to be made in the construction of biological theory. After all, life might have turned out differently; its fundamental organization might have differed from modern cells in its earliest stages billions of years ago, and it might be different on other planets or in other senses (e.g. artificial life). One important lesson of chemoton theory is that the sort of control exerted by nucleotide-based genes in modern cells must be an emergent property of chemoton organization and thus is the kind of property that should be investigated as an empirical result of the operation of evolutionary processes, not assumed as an axiom of biological explanation and theory. In Gánti's model of the simple chemoton, the 'genetic molecule' is a polymer made from a *single* kind of monomer; thus it *cannot* have a centralized control of function by virtue of the specificity of its base sequence, because its base sequence is merely V–V–V– $\cdots$.

These genetic polymers are stoichiometrically coupled in the chemoton model to both the autocatalytic central metabolism and the autocatalytic membrane. Because of this coupling, the genetic polymer can carry 'excess information', i.e. information about the states of the other chemical subsystems of the chemoton as well as about the state of the chemoton as a whole. The genetic molecule grows at a rate proportional to the number of cycles of the metabolism and proportional to the size of the membrane. Hence the genetic molecule stores information about the developmental state of the chemoton—how close it is to dividing/reproducing.

Stoichiometry thus leads to a *particular* way of understanding control functions in terms of three kinds of chemical properties: reaction network topologies, reaction specificities, and reaction rates. Having a control function depends on being upstream of other stages of a process. This topological dependency of control functions has been broadly investigated philosophically under the label of 'generative entrenchment' (Schank and Wimsatt 1988; Wimsatt and Schank 1988; Wimsatt 1999, in press). Further, both the specificity with which a particular kind of molecule undergoes a given reaction in particular conditions and the rate at which that reaction occurs are aspects of the way in which a particular reaction can exert control over a system (network) of reactions. Reaction specificities determine the chemical nature of the network constituents and, in conjunction with exogenous nutrients, the network topology. Reaction rates determine the relative concentrations and dynamics of the operation of the system of reactions. *Central* control, the kind attributed to DNA on the basis of its spatial geometrical centrality in the modern cell, is dynamic control by subsystems that are stoichiometrically intermediate to a system of reactions in a network, i.e. they are upstream of at least some other subsystems but there are chemical feedbacks to them as well. Thus the chemoton organization describes a very basic and fundamental condition for genetic molecules to be central controllers: they are intermediates in the chemical operating of the chemoton and they exert control.

This is an appealing result because it is consistent with contemporary philosophical resistance to the unique master role of the DNA, but at the same time leaves open the empirical possibility that DNA really does play the master role that the central dogma of molecular genetics implies. Stoichiometric control can be distributed *throughout* the chemical super-system of reactions, and yet it can make sense to distinguish the character of the control exerted by one type of controller (e.g. genetic molecule) from that of another (e.g. membrane).

The developmental systems perspective in the philosophy of biology extends the claim of distributed control beyond the bounds of the organism *per se* to include 'external' environments as integral parts of developmental systems (Oyama 1985; Gray 1992; Moss 1992; Griffiths and Gray 1994; Neumann-Held 1998; Oyama *et al.* 2001); on the distinction between developmental systems perspective and developmental systems theory, see Griesemer (2000b). Developmental systems theory argues that whole developmental systems are the 'replicators' or units of inheritance (Griffiths and Gray 1994). What chemoton theory adds to this distributed perspective is a set of specifically *chemical* notions of control implicit in stoichiometry and mass-balance equations and how to specify them. Thus it adheres to the traditional belief that there is *something* special about DNA's developmental and hereditary role, namely its sequence-based coding function (Godfrey-Smith 2000). At the same time, it facilitates investigations of distributed control which do not adhere to the master molecule model and without endorsing the conceptual strategy extending outward from genes to include other possible replicators (Sterelny *et al*. 1996), which risks further embedding of problematic assumptions about modern genes in the general analysis of replicators. Indeed, DNA-based genes are probably latecomers to the evolutionary replicator game from the point of view of chemoton evolution. Philosophers should avoid 'nucleic acid bias' in their understanding of life's organization just as much as they should avoid 'vertebrate bias' in their understanding of biological species.

One limitation of developmental systems theory has been its paucity of tools for delimiting developmental systems and thus its ability to make clear the role played by a given developmental 'resource', whether inside or outside a body, in structuring a system's developmental capacities, hereditary tendencies, and evolutionary potential. As critique, there is much to sympathize with in the developmental systems perspective, but there is much work to do before it can become a functioning alternative research programme. Gánti's cycle stoichiometry is a useful heuristic tool in this regard because it both constrains and guides exploration of control and regulatory relations and, more generally, interactions among components within a specified model of living organization. The constraint to satisfy the mass-balance requirements of stoichiometry could provide useful specificity for at least some models of developmental systems. It seems likely that other systems, such as symbolic inheritance systems, function precisely because they violate the conservation laws implicit in stoichiometry (Lachmann *et al*. 2000).

5.5.2   Stoichiometry disciplines

Stoichiometry adds a measure of discipline to philosophical efforts to formulate general accounts of units of heredity, development and evolution. The history of twentieth-century biology, and the dominant role of classical and molecular genetics within that history, have made it difficult even to think of other ways of describing biology than in terms of genes and their effects. Another dominant tradition of twentieth-century biology has been reductionism, which includes the view that principles, laws, and theories of phenomena at high levels of spatial organization (like organisms) must be consistent with principles, laws and theories of phenomena at lower levels of spatial organization (like atoms and molecules). This consistency is a necessary condition to satisfy the reductionist's dream of reducing biological theories to physical theories, and holds even for those of us who consider the real value of reductionism to lie in its heuristic aspects (Wimsatt 1976, 1987; Griesemer 2000a). It is a virtue of Gánti's cycle stoichiometry that it is applicable to life processes at their most fundamental, but is utterly indifferent to our biological bias toward studying and explaining life in terms of particular classes of molecules and molecular processes. *All* the molecules of a chemical system must be considered to satisfy the rules of stoichiometry. Thus, if control over the course of a developing chemical system can be exerted by a simple 'environmental' nutrient or stimulus, stoichiometry does not care whether that molecule is a DNA molecule in another body (as Dawkins (1982) examined in *The Extended Phenotype*) or a metal ion released from a mining operation to float down a contaminated river and be ingested by a fish. In organizational terms, stoichiometry provides the discipline to follow the molecular networks wherever they may lead rather than the organizational charts sanctioned by molecular dogmas describing how an organization is 'supposed' to work. The chemoton model counsels us to 'follow the stuff' as it moves around the chemical network and, in following the stuff, we will come to a clearer idea what role information of various sorts can and does play in that circulation of matter. 'Life', as Gánti begins his story, 'is an organization bound to matter'.

5.6 False models as means to truer theories

Traditionally, most biologists have been hostile to the goals and methods of 'theoretical biology'. Scepticism seems to flow from the attention that theoreticians often give to models obviously and blatantly simplified to the point that nothing stands out about them other than their falsity and even absurdity. Physicists have tended to be more tolerant of extreme simplification in their theoretical representations, perhaps because they have also tended towards equally extreme measures in their laboratories to produce, simplify, and control phenomena. Biologists have tended not to be as willing to force nature to fit procrustean beds of theory, laboratory walls, and existing technology (or at least not as willing to admit when they are doing so). This has led some to characterize biology as a different and more experiment-driven kind of science than physics, or at least to claim that biology has a different relationship to its theories than do the physical sciences. Thus one might expect even greater animus towards Gánti's vision of an 'exact theoretical biology'. Exactness means absolute precision, and Gánti's hope of finding absolute precision in biology is surely a pipe-dream.

Let us not misunderstand the goal and strategy of the exact theoretical biology that Gánti calls for. The aim is to produce an exact and complete *model* of the living state—an *abstraction* admitting of absolute precision in the *formal* specification of the living, not an absolutely precise *empirical* fit to the actual world of living things. The chemoton is an exact model in the sense that the calculations that can be made to predict and describe its behaviour are exactly determinable. This does not mean that anyone should expect to find chemotons in nature or that chemoton theory should lead directly to empirically accurate quantitative predictions of the behaviour of modern cells or organisms. Gánti is explicit on this point—the chemoton is an abstract model with a mainly heuristic function. It is a false model that can nevertheless be an effective tool in making predictions and generating explanations about phenomena (Wimsatt 1987).

Gánti's contributions to science can best be understood in terms of his background in chemical engineering, leading to his penchant to view models as tools for constructive theoretical science rather than as accurate representations of nature. The scientific value of a model lies as much in what one can *do* with it as in what one can *say* about it. Indeed, it has long been known that accuracy in *representation* is only one virtue among many that a theoretical model in biology might have (Levins 1966). The philosophical problem raised by the use of false models is to understand how and why they are used at all, given that the goal of science is to aim at the truth about nature (Wimsatt 1987). Gánti's work shows that the principal use of theoretical models like the chemoton is constructive rather than representational; they provide a framework for constructing the principles which can then guide representational efforts. They are tools to calculate *with*, not only to calculate *from*. If Levins is right that a model is 'a reconstruction of nature for the purpose of study' (Levins 1968, p. 6), then we should pay as much attention to purposes, ways, and means of study as we do to the nature of and empirical evidence for representations. Gánti's chemoton model is not intended to be true in the sense of an accurate representation of living cells.

Nor are Gánti's life criteria designed to stand on their own. They require the chemoton model to complete the specification of the living state because the model and its cycle stoichiometry supply a characterization of the *non*-living units out of which things in the living state are composed. Gánti's analysis of the living state is not to be identified with his life criteria, real or potential, but with these criteria in conjunction with the chemoton as their working model. Likewise, the chemoton is a hypothetical construct, not a complete model of the universal or primordial living cell because it could not actually exist. Actual cells have many properties and parts besides those of a simple chemoton.

The chemoton's functioning as a model depends on the life criteria to give it engineering 'blueprint sense' and the life criteria must be completed by the chemoton and cycle stoichiometry. This integrative approach to criteria and models represents a philosophically sophisticated position. It is in harmony with philosophical interest over the last 30 years in the role of models in the actual practice of science. The ideas presented in Gánti's work introduce us to the coherent and integrated picture of what Gánti calls 'exact theoretical biology', a picture of the very science that many philosophers of evolution seek.

References

Abir-Am, P. (1985). Themes, genres and orders of legitimation in the consolidation of new scientific disciplines: deconstructing the historiography of molecular biology. *History of Science*, **23**, 73–117.

Alberts, B., Johnson, A., Lewis, J., Raff, M., Roberts, K., and Walter, P. (1994). *Molecular biology of the cell* (3rd edn). Garland, New York.

Ashby, W.R. (1952). *Design for a brain*. Wiley, New York.

Bachmann, P.A., Luisi, L., and Lang, J. (1992). Autocatalytic self-replicating micelles as models for prebiotic structures. *Nature*, **357**, 57–9.

Barrow, J.D. and Tipler, F.J. (1988). *The anthropic cosmological principle*. Clarendon Press, Oxford.

Beatty, J. (1981). What's wrong with the received view of evolutionary theory? In *PSA 1980*, Vol. 2, pp. 397–426. Philosophy of Science Association, East Lansing, MI.

Benner, S.A. *et al.* (1987). Natural selection, protein engineering, and the last ribo-organism: rational model building in biochemistry. *Cold Spring Harbor Symposia on Quantitative Biology*, **52**, 56–63.

Benner, S.A., Ellington, D.A., and Tauer, A. (1989). Modern metabolism as a palimpsest of the RNA world. *Proceedings of the National Academy of Sciences of the United States of America*, **86**, 7054–8.

Brandon, R. (1990). *Adaptation and environment*. Princeton University Press.

Buss, L.W. (1983). Evolution, development, and the units of selection. *Proceedings of the National Academy of Sciences of the United States of America*, **80**, 1387–91.

Buss, L.W. (1987). *The evolution of individuality*. Princeton University Press.

Cahill, D., Kinzler, K.W., Vogelstein, B., and Lengauer, C. (1999). *Trends in Genetics*, **15**, M57–60.

Cartwright, N. (1983). *How the laws of physics lie*. Clarendon Press, Oxford.

Cartwright, N. (1989). *Nature's capacities and their measurement*. Clarendon Press, Oxford.

Cavalier-Smith, T. (1995). Membrane heredity, symbiogenesis, and the multiple origins of algae. In *Biodiversity and evolution* (ed. R. Arai, M. Kato, and Y. Doi), pp. 75–114. National Science Museum Foundation, Tokyo.

Clark, W.R. (1996). *Sex and the origins of death*. Oxford University Press.

Collier, J.D. and Hooker, C.A. (1999). Complexly organized dynamical systems. *Open Systems and Information Dynamics*, **6**, 241–302.

Covert, M.W., Schilling, C.H., Famili, J.S., *et al.* (2001). Metabolic modeling of microbial strains *in silico*. *Trends in Biochemical Science*, **26**, 179–86.

Crick, F.H.C. (1968). The origin of the genetic code. *Journal of Molecular Biology*, **38**, 367–79.

Crick, F.H.C. (1981). *Life itself: its origin and nature*. Simon and Schuster, New York.

Dawkins, R. (1976). *The selfish gene*. Oxford University Press.

Dawkins, R. (1982). *The extended phenotype: the gene as a unit of selection*. Oxford University Press.

Deamer, D.W. (1998). Membrane compartments in prebiotic evolution. In *The molecular origins of life* (ed. A. Brack). pp. 189–205. Cambridge University Press.

Deamer, D.W. and Fleischaker, G.R. (ed.) (1994). *Origins of life: the central concepts.* Jones and Bartlett, Boston, MA.

De Duve, C. (1991). *Blueprint for a cell: the nature and origin of life.* Neil Patterson, Burlington, NC.

Dobzhansky, T. (1970). *Genetics of the evolutionary process.* Columbia University Press, New York.

Dretske, F. (1981). *Knowledge and the flow of information.* MIT Press, Cambridge, MA.

Eigen, M. (1971). Self-organization of matter and the evolution of biological macromolecules. *Naturwissenschaften*, **58**, 465–523.

Eigen, M. and Winkler-Oswatitsch, R. (1996). *Steps toward life: a perspective on evolution* (trans. P. Woolley). Oxford University Press.

Feldman, F. (1992). *Confrontations with the reaper: a philosophical study of the nature and value of death.* Oxford University Press, New York.

Ferris, J.P., Hill, A.R., Liu, R., and Orgel, L.E. (1996). Synthesis of long prebiotic oligomers on mineral surfaces. *Nature*, **381**, 59–61.

Fontana, W. and Buss, L.W. (1994a). What would be conserved if 'the tape were played twice'? *Proceedings of the National Academy of Sciences of the United States of America*, **91**, 757–61.

Fontana, W. and Buss, L.W. (1994b). 'The arrival of the fittest': toward a theory of biological organization. *Bulletin of Mathematical Biology*, **56**, 1–64.

Fontana, W., Wagner, G., and Buss, L.W. (1994). Beyond digital naturalism. *Artificial Life*, **1**, 211–27.

Fox, S.W. (1965). A theory of macromolecular and cellular origins. *Nature*, **205**, 328–40.

Gánti, T. (1971). *Az élet princípuma (The principle of life).* Gondolat, Budapest (in Hungarian).

Gánti, T. (1974). Theoretical deduction of the function and structure of the genetic material. *Biológia*, **22**, 17–35.

Gánti, T. (1979a). Interpretation of prebiotic evolution on the basis of chemoton theory. *Biológia*, **27**, 161–75 (in Hungarian).

Gánti, T. (1979b). *A theory of biochemical supersystems and its application to problems of natural and artificial biogenesis.* University Park Press, Baltimore, MD.

Gánti, T. (1984). *Chemoton theory*, Vol I. OMIKK, Budapest (in Hungarian).

Gánti, T. (1987). *The principle of life* (6th edn). OMIKK, Budapest (first published in Hungarian in 1971).

Gánti, T. (1989a). *Chemoton theory*, Vol II. OMIKK, Budapest (in Hungarian).

Gánti, T. (1989b). *Contra Crick, or the nature of life* (in Hungarian). Gondolat, Budapest.

Gánti, T. (1991). *Chemoton theory* (trans. E. Czárán). Vol I: *Theoretical foundation of fluid machineries and chemoton theory.* Vol. II: *Theory of living systems.* OMIKK, Budapest.

Gánti, T. (1997). Biogenesis itself. *Journal of Theoretical Biology*, **187**, 583–93.

Gánti, T. (2000). *The general theory of life* (in Hungarian). Müszaki Könyvkiadó, Budapest.

Giere, R. (1988). *Explaining science: a cognitive approach.* University of Chicago Press.

Giere, R. (1997). *Understanding scientific reasoning* (4th edn). Harcourt Brace College Publishers, Fort Worth, TX.

Gilbert, W. (1986). The RNA world. *Nature*, **319**, 818.

Godfrey-Smith, P. (2000). On the theoretical role of 'genetic coding'. *Philosophy of Science*, **67**, 26–44.

Gray, R. (1992). Death of the gene: developmental systems strike back. In *Trees of life* (ed. P.E. Griffiths). Kluwer Academic, Dordrecht.

Grey, D., Hutson, V., and Szathmáry, E. (1995). A re-examination of the stochastic corrector model. *Proceedings of the Royal Society of London, Series B*, **262**, 29–35.

Griesemer, J.R. (1996a). Some concepts of historical science. *Memorie della Societá Italiana de Scienze Naturali e del Museo Civico di Storia Naturale di Milano*, **27**, 60–9.

Griesemer, J.R. (1996b). Periodization and models in historical biology. In *New perspectives on the history of life*, pp. 19–30. California Academy of Sciences, San Francisco, CA.

Griesemer, J.R. (1998). Commentary: the case for epigenetic inheritance in evolution. *Journal of Evolutionary Biology*, **11**, 193–200.

Griesemer, J.R. (2000a). Reproduction and the reduction of genetics. In *The concept of the gene in development and evolution: historical and epistemological perspectives* (ed. P. Beurton, R. Falk, and H.-J. Rheinberger. Cambridge University Press.

Griesemer, J.R. (2000b). Development, culture and the units of inheritance. *Philosophy of Science*, **67**, S348–68.

Griesemer, J.R. (2000c). The units of evolutionary transition. *Selection*, **1**, 67–80.

Griffiths, P.E. and Gray, R. (1994). Developmental systems and evolutionary explanation. *Journal of Philosophy*, **91**, 277–304.

Griffiths, P.E. and Neumann-Held, E. (1999). The many faces of the gene. *BioScience*, **49**, 656–62.

Hull, D.L. (1981). The units of evolution: a metaphysical essay. In *The philosophy of evolution* (ed. U. Jensen and R. Harré). Harvester Press, Brighton.

Hull, D.L. (1988). *Science as a process*. University of Chicago Press.

Hurst, L.D., Atlan, A., and Bengtsson., B.O. (1996). Genetic conflicts. *Quarterly Review of Biology*, **71**, 317–64.

Hutchison, C.A., Peterson, S.N., Gill, S.R., *et al.* (1999). Global transposon mutagenesis and minimal plasma genome. *Science*, **286**, 2165–9.

Huxley, J. (1912). *The individual in the animal kingdom*. Cambridge University Press.

Jablonka, E. and Lamb, M. (1995). *Epigenetic inheritance and evolution*. Oxford University Press.

Jablonka, E. and Szathmáry, E. (1995). The evolution of information storage and heredity. *Trends in Ecology and Evolution*, **10**, 206–11.

Jackson, J., Buss, L., and Cook, R. (ed.) (1985). *Population biology and evolution of clonal organisms*. Yale University Press, New Haven, CT.

Johnstone, W.K., Unrau, P.J., Lawrence, M.S., Glasner, M.E., and Bartel, D.P. (2001). RNA-catalyzed RNA polymerization: accurate and general RNA-plated primer extension. *Science*, **292**, 1319–25.

Joyce, G. (1994). Foreword. In *Origins of life: the central concepts* (ed. D.W. Deamer and G.R. Fleischaker). Jones and Bartlett, Boston, MA.

Joyce, G. and Orgel, L.E. (1999). Prospects for understanding the origin of the RNA world. In *The RNA world* (2nd edn) (ed. F. Gesteland, T.R. Cech, and J.F. Atkins), pp. 49–77. Cold Spring Harbor Laboratory, Cold Spring Harbor, NY.

Jubien, M. (1997). *Contemporary metaphysics*. Blackwell, Oxford.

Judson, H.F. (1979). *The eighth day of creation: the makers of the revolution in biology*. Simon and Schuster, New York.

Kauffman, S.A. (1986). Autocatalytic sets of proteins. *Journal of Theoretical Biology*, **119**, 1–24.

Kay, L. (1993). *The molecular vision of life: Caltech, the Rockefeller Foundation, and the rise of the new biology*. Oxford University Press, New York.

Kay, L. (2000). *Who wrote the book of life? A history of the genetic code*. Stanford University Press.

Kemeny, J. (1955). Man viewed as a machine. *Scientific American*, **152**, 58–67.

King, G.A.M. (1982). Recycling, reproduction, and life's origin. *BioSystems*, **15**, 89–97.

King, G.A.M. (1986). Was there a prebiotic soup? *Journal of Theoretical Biology*, **123**, 493–8.

Kirkwood, T.B.L. and Austad, S.N. (2000). Why do we age? *Nature*, **408**, 233–8.

Kuhn, T.S. (1970). *The structure of scientific revolutions* (2nd edn). University of Chicago Press.

Lachmann, M., Sella, G., and Jablonka, E. (2000). On the advantages of information sharing. *Proceedings of the Royal Society of London, B*, **267**, 1287–93.

Lenski, R.E., Ofria, C., Collier, T.C., and Adami, C. (1999). Genome complexity, robustness and genetic interactions in digital organisms. *Nature*, **400**, 661–4.

Levins, R. (1966). The strategy of model building in population biology. *American Scientist*, **54**, 421–31.

Levins, R. (1968). *Evolution in changing environments: some theoretical explorations*. Monographs in Population Biology No. 2, Princeton University Press.

Lewontin, R.C. (1970). The units of selection. *Annual Review of Ecology and Systematics*, **1**, 1–17.

Lewontin, R.C. (1978). Adaptation. *Scientific American*, **239**, 213–30.

Lewontin, R.C. (1983). Gene, organism and environment. In *Evolution from molecules to men* (ed. D.S. Bendall), pp. 273–85. Cambridge University Press.

Lewontin, R.C. and Dunn, L.C. (1966). The evolutionary dynamics of a polymorphism in the house mouse. *Genetics*, **45**, 705–22.

Lloyd, E.S. (1988). *The structure and confirmation of evolutionary theory*. Greenwood Press, New York.

Luisi, P.L. (1998). About various definitions of life. *Origins of Life and Evolution of the Biosphere*, **28**, 613–22.

Luisi, P.L., Walde, P., and Oberholzer, T. (1994). Enzymatic RNA synthesis in self-reproducing vesicles: an approach to the construction of a minimal synthetic cell. *Berichte der Bunsengesellschaft für Physikalische Chemie*, **98**, 1160–5.

Luther, A., Brandsch, R., and von Kiedrowski, G. (1998). Surface-promoted replication and exponential amplification of DNA analogues. *Nature*, **396**, 245–8.

Mahner, M. and Bunge, M. (1997). *Foundations of biophilosophy*. Springer-Verlag, Berlin.

Margulis, L. (1981). *Symbiosis in cell evolution: life and its environment on the early Earth*. W.H. Freeman, San Francisco, CA.

Maturana, H.R. (1981). The organization of the living: a theory of the living organization. In *Cybernetics Forum*, **10** (special issue devoted to autopoiesis, ed. M. Zeleny). (Originally published in *International Journal of Man–Machine Studies*, **7**, 1975).

Maturana, H.R. and Varela, F.J. (1980). *Autopoiesis and cognition: the realization of the living*. Reidel, Dordrecht.

Maynard Smith, J. (1986). *The problems of biology*. Oxford University Press.

Maynard Smith, J. (1987). How to model evolution. In *The latest on the best: essays on evolution and optimality* (ed. J. Dupré), pp. 119–31. MIT Press, Cambridge, MA.

Maynard Smith, J. (2000). The concept of information in biology. *Philosophy of Science*, **67**, 177–94.

Maynard Smith, J. and Szathmáry, E. (1995). *The major transitions in evolution*. Oxford University Press.

Maynard Smith, J. and Szathmáry, E. (1999). *The origins of life: from the birth of life to the origin of language*. Oxford University Press.

Morgan, M.S. and Morrison, M. (ed.) (1999). *Models as mediators: perspectives on natural and social sciences*. Cambridge University Press.

Morowitz, H.J. (1979). The six million dollar man. In *The Wine of Life*, pp. 3–6. St. Martin's Press, New York.

Morowitz, H.J. (1992). *Beginnings of cellular life: metabolism recapitulates biogenesis*. Yale University Press, New Haven, CT.

Morowitz, H.J. and Deamer, D.W. (1988). The chemical logic of a minimum protocell. *Origins of Life and Evolution of the Biosphere*, **18**, 281–7. (Reprinted in *Origins of life: the central concepts* (ed. D.W. Deamer and G.R. Fleischaker), pp. 263–9. Jones and Bartlett, Boston, MA, 1994.)

Moss, L. (1992). A kernel of truth? On the reality of the genetic program. In *Proceedings of the Philosophy of Science Association*, Vol. 1 (ed. D. Hull, M. Forbes, and K. Okruhlik). Philosophy of Science Association, East Lansing MI, 335–48.

Müller, D., *et al.* (1990). Chemie von alpha-Aminonitrilen. *Helvetia Chimica Acta*, **73**, 1410–68.

Muller, H.J. (1922). Variation due to change in the individual gene. *American Naturalist*, **56**, 32–50.

Mushegian, A.R. and Koonin, E.V. (1996). A minimal gene set for cellular life derived by comparison of complete bacterial genomes. *Proceedings of the National Academy of Sciences of the United States of America*, **93**, 10 268–73.

Neumann-Held, E. (1998). The gene is dead—long live the gene! Conceptualizing genes the constructionist way. In *Sociobiology and bioeconomics: the theory of evolution in biological and economic theory* (ed. P. Koslowski). Springer-Verlag, New York.

Nitecki, M.H. and Nitecki, D.V. (ed.) (1992). *History and evolution*. State University of New York Press, Albany, NY.

Odling-Smee, F.J., Laland, K.N., and Feldman, M.W. (1996). Niche construction. *American Naturalist*, **147**, 641–8.

Olby, R. (1974). *The path to the double helix*. University of Washington Press, Seattle, WA.

Oparin, A.I. (1961). *Life: its nature, origin and development*. Academic Press, New York.

Orgel, L.E. (1968). Evolution of the genetic apparatus. *Journal of Molecular Biology*, **38**, 381–93.

Orgel, L.E. (1990). Adding to the genetic alphabet. *Nature*, **343,** 18–20.

Orgel, L.E. (2000). Self-organizing biochemical cycles. *Proceedings of the National Academy of Sciences of the United States of America*, **97**, 12 503–7.

Oyama, S. (1985). *The ontogeny of information: developmental systems and evolution*. Cambridge University Press.

Oyama, S. (2000). *The ontogeny of information: developmental systems and evolution (science and cultural theory)*. Duke University Press, Durham, NC.

Oyama, S., Griffiths, P.E., and Gray, R. (ed.) (2001). *Cycles of contingency: developmental systems and evolution*. MIT Press, Cambridge, MA.

Penrose, L.S. (1959). Self-reproducing machines. *Scientific American*, **200**, 105–14.

Piccirilli, J.A., Krauch, T., Moroney, S.E., and Benner, S.A. (1990). Enzymatic incorporation of a new base pair into DNA and RNA extends the genetic alphabet. *Nature*, **343**, 33–7.

Pierce, C.S. (1897). Logic as semiotic: the theory of signs. Reprinted in *Philosophical writings of Pierce* (ed. J. Buchler). Dover Publications, New York, 1955.

Rashevsky, N. (1938). *Mathematical biophysics.* University of Chicago Press.

Rebek, J. (1994). Synthetic self-replicating molecules. *Scientific American*, **271**, 48–53, 55.

Robert, J.S. (2001). Interpreting the homeobox: metaphors of gene action and activation in development and evolution. *Evolution and Development*, **3**, 287–95.

Rössler, O. (1971). Ein systemtheoretisches Modell zur Biogenese. (A system theoretic model of biogenesis.) *Zeitschrift für Naturforschung*, **26b**, 741–6.

Rössler, O. (1972). Grundschaltungen von flüssigen Automaten und Relaxationssystemen. (Basic switches of fluid automata and relaxation systems.) *Zeitschrift für Naturforschung*, **27b**, 333–43.

Rössler, O. (1974). Chemical automata in homogeneous and reaction–diffusion kinetics. In *Physics and mathematics of the nervous system* (ed. M. Conrad *et al.*), pp. 399–418. Springer-Verlag, Berlin.

Santelices, B. (1999). How many kinds of individual are there? *Trends in Ecology and Evolution*, **14**, 152–5.

Sapp, J. (1987). *Beyond the gene: cytoplasmic inheritance and the struggle for authority in genetics.* Oxford University Press, New York.

Schank, J.C. and Wimsatt, W.C. (1988). Generative entrenchment and evolution. In *PSA-1986* (ed. A. Fine and P.K. Machamer). Philosophy of Science Association, East Lansing, MI.

Schilling, C.H. and Palsson, B.O. (1998). The underlying pathway structure of biochemical reaction networks. *Proceedings of the National Academy of Sciences of the United States of America*, **95**, 4193–8.

Schrödinger, E. (1944). *What is life? The physical aspect of the living cell.* Cambridge University Press.

Segré, D., Ben-Eli, D., Deamer, D.W., and Lancet, D. (2001). The lipid world. *Origins of Life and Evolution of the Biosphere*, **31**, 119–45.

Shapiro, R. (1987). *Origins: A skeptic's guide to the creation of life on Earth.* Bantam Doubleday, New York.

Sharon, N. and Lis, H. (1993). Carbohydrates in cell recognition. *Scientific American*, **268**, 74–81.

Simon, H. (1962). The architecture of complexity. Reprinted in *The sciences of the artificial*, MIT Press, Cambridge, MA, 1981.

Sober, E. (1984). *The nature of selection: evolutionary theory in philosophical focus.* MIT Press, Cambridge, MA.

Sober, E. and Wilson, D.S. (1998). *Unto others: the evolution and psychology of unselfish behavior.* Harvard University Press, Cambridge, MA.

Spiegelman, S. (1970). Extracellular evolution of replicating RNA molecules. In *The neurosciences: second study program* (ed. F.O. Schmitt), pp. 927–45. Rockefeller University Press, New York.

Sterelny, K., Smith, K., and Dickison, M. (1996). The extended replicator. *Biology and Philosophy*, **11**, 377–403.

Strobel, S.A. (2001). Repopulating the RNA world. *Nature*, **411**, 1003–6.

Szathmáry, E. (1992a). Natural selection and the dynamical coexistence of defective and complementing virus segments. *Journal of Theoretical Biology*, **157**, 383–406.

Szathmáry, E. (1992b). What determines the size of the genetic alphabet? *Proceedings of the National Academy of Sciences of the United States of America*, **89**, 2614–18.

Szathmáry, E. (1993a). Molecular variation and the evolution of viruses. *Trends in Ecology and Evolution*, **8**, 8–9.

Szathmáry, E. (1993b). Coding coenzyme handles: a hypothesis for the origin of the genetic code. *Proceedings of the National Academy of Sciences of the United States of America*, **90**, 9916–20.

Szathmáry, E. (1998). Useful stuff. *Trends in Ecology and Evolution*, **13**, 251–2.

Szathmáry, E. (2000). The evolution of replicators. *Philosophical Transactions of the Royal Society of London. Series B, Biological Sciences*, **355**, 1669–76.

Szathmáry, E. (2001a) Biological information, kin selection, and evolutionary transitions. *Theoretical Population Biology*, **59**, 11–14.

Szathmáry, E. (2001b). Evolution. Developmental circuits rewired. *Nature*, **411**, 143–5.

Szathmáry, E. and Demeter, L. (1987). Group selection of early replicators and the origin of life. *Journal of Theoretical Biology*, **128**, 463–86.

Szathmáry, E. and Maynard Smith, J. (1993). The origin of genetic systems. *Abstracta Botanica*, **17**, 197–206.

Szathmáry, E. and Maynard Smith, J. (1995). The major evolutionary transitions. *Nature*, **374**, 227–32.

Szathmáry, E. and Maynard Smith, J. (1997). From replicators to reproducers: the first major transitions leading to life. *Journal of Theoretical Biology*, **187**, 555–72.

Szostak, J.W., Bartel, D.P., and Luisi, P.L. (2001). Synthesizing life. *Nature*, **409**, 387–90.

Tarumi, K. and Schwegler, H. (1987). A non-linear treatment of the protocell model by a boundary layer approximation. *Bulletin of Mathematical Biology*, **47**, 307–20.

Timoféev-Ressovsky, N., Zimmer, K., and Delbrück, M. (1935). über die natur der Genmutation und der Genstruktur. *Nachrichten von der Gesellschaft der Wissenschaften zu Göttingen, Mathematisch-Physikaliche Klasse*, **6**, 189–245.

Turner, P.E. and Chao, L. (1999). Prisoner's dilemma in an RNA virus. *Nature*, **398**, 441–3.

van Fraassen, B. (1980). *The scientific image*. Clarendon Press, Oxford.

Varela, F.G. (1979). *Principles of biological autonomy*. North-Holland, New York.

Varela, F.G., Maturana, H.R., and Uribe, R. (1981). Autopoiesis: the organization of living systems, its characterization and a model. In *Cybernetics Forum*, **10** (special issue devoted to autopoiesis, ed. M. Zeleny). (Originally published in *BioSystems*, **5**, 187–96, 1974.)

von Kiedrowski, G. (1986). A self-replicating hexadeoxy nucleotide. *Angewandte Chemie International Edition in English*, **25**, 932–5.

von Kiedrowski, G. (1996). Primordial soup or crêpes? *Nature*, **381**, 20–1.

von Neumann, J. and Burks, A.W. (1966). *The theory of self-reproducing automata*. University of Illinois Press, Urbana, IL.

Wächtershäuser, G. (1988). Before enzymes and templates: theory of surface metabolism. *Microbiology Review*, **52**, 452–84.

Wächtershäuser, G. (1992). Groundwork for an evolutionary biochemistry: the iron–sulfur world. *Progress in Biophysics and Molecular Biology*, **58**, 85–201.

Waddington, C.H. (1975). *The evolution of an evolutionist*. Cornell University Press, Ithaca, NY.

Walde, P.W. *et al.* (1994). Autopoietic self-reproduction of fatty acid vesicles. *Journal of the American Chemical Society*, **116**, 11 649–54.

Watson, J.D. (1992). A personal view of the project. In *The code of codes: scientific and social issues in the Human Genome Project* (ed. D. Kevles and L. Hood), pp. 164–73. Harvard University Press, Cambridge, MA.

White, H.B. (1976). Coenzymes as fossils of an earlier metabolic stage. *Journal of Molecular Evolution*, **7**, 101–4.

Williams, G.C. (1992). *Natural selection: domains, levels, and challenges*. Oxford University Press.

Wimsatt, W.C. (1974). Complexity and organization. In *PSA-1972* (*Boston Studies in the Philosophy of Science*, Vol. 20), pp. 67–86. Reidel, Dordrecht.

Wimsatt, W.C. (1976). Reductive explanation: a functional account. In *PSA 1974* (ed. A.C. Michalos, C.A. Hooker, G. Pearce, and R.S. Cohen). Reidel, Boston, MA.

Wimsatt, W.C. (1980). Reductionist research strategies and their biases in units of selection controversy. In *Scientific discovery: historical and scientific case studies* (ed. T. Nickles). Reidel, Dordrecht.

Wimsatt, W.C. (1985). Forms of aggregativity. In *Human nature and natural knowledge* (ed. A. Donagan, A. Perovich, and M. Wedin). Reidel, Dordrecht.

Wimsatt, W.C. (1987). False models as means to truer theories. In *Neutral models in biology* (ed. M. Nitecki and A. Hoffman). Oxford University Press.

Wimsatt, W.C. (1997). Aggregativity: reductive heuristics for finding emergence. *Philosophy of Science*, **64** (Supplement 4), S372–84.

Wimsatt, W.C. (1999). Generativity, entrenchment, evolution, and innateness. In *Biology meets psychology: philosophical essays* (ed. V. Hardcastle). MIT Press, Cambridge, MA.

Wimsatt, W.C. *Re-engineering philosophy for limited beings: piecewise approximations to reality*. Harvard University Press, Cambridge, MA, in preparation.

Woese, C.R. (1967). *The genetic code*. Harper & Row, New York.

Wolfram, S. (1994). *Cellular automata and complexity*. Perseus Books, New York.

Wright, E.O., Levine, A., and Sober, E. (1992). *Reconstructing Marxism: essays on explanation and the theory of history*. Verso, London.

Yoder, J.A., Walsh, C.P., and Bestor, T.H. (1997). Cytosine methylation and the ecology of intragenomic parasites. *Trends in Genetics*, **12**, 335–40.

Index

abiogenesis 133, 134, 141–2
abiotic loop RNA formation/
    combination 47–9, 51
  synthesizing enzyme assembly 52
abiotic nucleic acid synthesis, failure
    of 140
absolute life criteria 76–8, 80, 82,
    106, 160–161, 162
  chemotons satisfying
    111–112, 114
  connection with potential life
    criteria 177
abstract chemoton networks 41
abstract constructions, chemotons
    as 38
abstract state fields 21, 22
acceleratory reactions 123
accumulation systems 73, 87,
    95–101, 148
acenaphthilene, polymerization
    of 109
acetyl coenzyme A 88, 97, 100, 146
acting principle 14
action chains 122, 123
adaptability, environmental,
    of chemotons 117
adenosine triphosphate *see* ATP
amino acids, experimental
    production of 134
ammonia and prebiotic
    chemoton 41, 42
ammonium cyanate 146
amplification of signs 123
ancestral cells 61
animal life 7, 8, 10
  vs automata 120–1
  vs plant life 9
ant hills as dynamic
    soft systems 65–6
aperiodic crystals 14, 15
approximation of reality 55, 56
Ashby's cybernetic stability
    criterion 73, 77, 86, 87
atmosphere
  composition 42, 134
  primordial 28, 39, 42, 54, 136–7
atomic energy liberation 155
ATP 26, 96, 98, 147
  in autocatalytic processes 26, 28,
    96, 146–7
  phosphorylation 96, 146, 147

autocatalysis/autocatalysts 4, 95–6,
    157, 167
autocatalytic processes
  ATP in 26, 28, 146–7
  cycles 96, 97, 98, 103–4, 109, 125,
    126, 138, 145
    coupling 27
    in microspheres 34
    non-enzymatic 26
    see also self-reproducing
      processes, cycles
  networks 44, 165
  see also reproduction
automata 3, 73, 148, 151, 160
  control of 13
  fluid 3–4, 16–19, 124, 175–8
    chemical 3, 18, 23, 30–1, 32
    primitive 32
    'ticking' 26–7
    proliferating 35
    'ticking' 24
  Leibniz on 12–13
  man-made *vs* living
    systems 12–13
  program-controlled 30–1, 32, 41
  self-reproducing 175
    fluid 124, 175–8
    von Neumann, J. on 124, 125
    without technical
      prescription 125
  soft 3, 120–4, 152
autonomy 176, 177
  biological 163
  genetic 8
autopoiesis 76, 78, 163–5, 176
  similarities to chemoton
    theory 177

back-mutation 153
Bahadur, K. on biogenesis 144–5
base pairs, alternative types 133
base sequences 13, 43, 44
  see also letter sequences
base substitution 129–30
basic units *see* units
Belousov–Zhabotinskii reaction 23,
    25, 27
binding sites
  secondary 48, 51

temporary 50
biochemical cycles 22, 87
biogenesis 11, 42, 50, 86,
    132, 133
  experimental 144–5, 156
  two-ribozyme system 168
  without genetic code 157
  see also life, origin of
biological clocks 124
biological manipulations, dangers
    of 152, 154
biological organization, hierarchy
    of 171
biological self-organization 176
biological systems 142, 177
  from simpler structures
    148, 149, 150
  unitary 66
biologically-important substances
  chemical evolution 134
  experimental formation 137
biologist, responsibility of
    152–6
biology 147
  as abstraction 57
  exobiology 80, 161
  and philosophy 169–75
  see also molecular biology;
    theoretical biology
bond strengths in chemotons
    130–1
brain 149, 150–1

Calvin cycle 22, 26, 96, 98,
    99, 146
  as connection between
    subsystems 146
  sugar phosphate production 34
carbon dioxide 41, 42
cells
  as dynamic soft systems 65
  metabolism 23
  subsystems 81–5
  three subsystem structure 83
cellular death 114
chain reactions 123
chemical bonds 130
chemical communication
  systems/signs 108

chemical cycles   22, 23, 28, 86, 96
  need for external energy   89–90
  stability   99–100
  and template
    polymerization   31–2
chemical energy   2–3, 18
chemical evolution   28, 29, 32, 33,
    39, 40, 42, 133, 135, 166
  compounds of biological
    importance   134
  experimental   29, 42
chemical incompatibility   40, 135
chemical kinetics   21, 32, 115, 165
chemical motor of cell   85–95, 100
  energy flow   94–5
  irreversibility in living
    systems   113
  'water wheel' analogy   95
chemical reactions
  cycles as constrained paths   74
  direction   19–20, 22
  energy flow   20
  non-enzymatic   136–7
  under primeval conditions   137
  rates   117–18, 127
chemical state fields *see* state fields,
  chemical
chemical supersystems   5, 85,
    106, 111
  proliferating   104
chemical systems
  autonomous evolution of   163
  effect of sensed changes   123
chemical work   71
  continuous   72
chemistry
  and abstract state space   21
  basic units   58, 67
  material balance equations   36
  metabolic processes   112
  total synthesis   143
chemoton(s)   3–6, 29–36, 40,
    107–11, 158
  as abstract constructions   38
  adaptability   44, 117
  as biological minimal system   85
  cell/life cycle   181
  cell/life cycles   162
  as cells   54, 150
  chemical construction   4
  connections   36, 38, 41, 110
    bond strengths   130–1
  division   38–9, 43, 119
  emergent properties   182
  and enzymes   53, 148
  exact numerical investigation   5
  formose reaction in   39

implications   1
  as inherent units   112
  and life criteria   112, 113–14, 162
  and living systems   38, 41
  as minimum biological system   1
  model(s)   4–6
    abstract/false   186
    bottom-up design   5
    chemical reaction rate   5
    concrete   150
    and eukaryotic cells   9
    exact   185
    minimum model   4
    mutative   129
    as representation of prokaryotic
      life   6–7
    and RNA   167
    three subsystem structure   179
  nutrients/waste products   41
  offspring cycles   176
  operation   37–8
  pacemaker function   140
  philosophical aspects   179–80
  prebiotic   36–42
  realization   38, 41
  RNA construction   40
  sensitivity   118
  and spatial separation   32
  stability   112, 119
  subsystems   36, 39, 110, 165
    compatibility of   145
    connectability   145
    controlling subunit   36–7
    experimental validation   138
    information subsystem   39–40,
      139–40
    membrane   37, 40–1, 43
    see also metabolic subsystem
  system organization   36
  three subsystem structure
    111, 165–6
chemoton-level life   111–15
chemoton theory   58
  applicability   57–8
  brain functional model from   151
  circularity   158, 173
  and exact science   57–8
  experimental validation   137–8
  and laws of genetics   124–32
  and living systems   134, 136, 142,
    143
    total synthesis   145
  and origin of life   132–40
  and principles of soft
    automata   120–4
  similarities to autopoiesis   177
  three subsystem structure   83

chloroplasts   61
citric acid cycle *see* Szent-
    Györgyi–Krebs cycle
clocks
  biological   124
  chemical   16–17, 123
closed action chains   123
comparative genomics   163
compartmentalization   162–3
computer simulations
  chemotons   114, 115–20
  'ticking' fluid automata   24
connection diagrams   24, 36, 38
  see also metabolic maps
connection system, abstract   41
constrained paths/motions   74, 82, 86
  closure to create stability   86–7
continuous work performance
  chemical   72
  conditions for   68
  and equilibrium   82
  Krebs cycle   90–1
control
  central   183
  distributed   182–4
  in living systems   78, 114
criteria of life *see* life criteria
crystallography   59, 66, 67
cyanopolymers   135
cybernetic stability criterion   86, 87
cycle stoichiometry   23, 32, 42, 164,
    178–9
  linking models to life criteria   182,
    186
  material balance equations   36
cytoplasm   6, 18, 83–4, 98
  accumulation   95
  as autocatalytic system   26
  as chemical automaton   18
  as chemical motor   19, 95
  as fluid automaton   17
  as homeostatic subsystem   99
  as machine   17
  as soft system   84, 101
cytoskeleton   17

Dawkins, R. on reproduction   79
death   2, 8, 75, 79–80
  of chemotons   114
definitions of life   160–2
  Gánti's view   162
developmental systems   8, 17,
    173, 178
  perspective   183–4
differential equations   115, 117, 118

distributed control 182–4
divisible systems 66, 111
  clocks 16
  organisms 68
division
  of chemotons 38–9, 43
  of microspheres 34–5, 44
DNA 13, 157
  in chemoton theory 183
  damage/repair 13
  material basis of heredity 125
  molecular structure 14–15, 131
  parasitic 13
  and replicator function 172
  sign sequences and enzymatic
    functions 148
  spontaneous formation 15
  synthesis 18, 109, 139
  Watson–Crick model of
    14–15, 45, 57, 124
double-molecule layers 102
double-strand state 131
dynamic stability 86
dynamic systems 64–5

Eigen, M. on origin of life 157, 158
electric charge 34
elementary cells 59, 67
endosymbiotic theory 9, 61, 62
energy
  chemical 2–3, 18
  in Krebs cycle general formula 94
  in mechanical *vs* living
    systems 2–3
  and monomolecular layers 101–2
  storage compounds 98
  transformation in living
    systems 18
energy flow 18
  chemical 20, 21, 94
  forced trajectories 19, 21
entelechia 14
enzymatic systems 128
  and chemoton theory 5
  controlling/regulating living
    cells 62, 63
enzyme(s) 45, 50, 130
  in abstract chemical state
    fields 22
  in chemistry 62–3
  'petrol engine' analogy 22
  in proteinoids 144
  regulation in chemoton
    theory 148
  single RNA strands 53

'water wheel' analogy 22
enzyme RNAs 50–1, 52–3
  see also ribozymes
epigenetic inheritance 13, 35
equilibria 73
  and living systems 72, 82
  as sensors 122
  for sets/systems 69, 71
equilibrium subsystem 153
eRNA *see* ribozymes
ethical issues 152–6
eukaryotic cells 9, 10, 61, 62, 149,
  150, 153
evolution 1, 3, 6, 10, 16, 35, 50, 53,
  125, 158, 161, 166, 170
  chemical *see* chemical evolution
  of chemotons 42, 45, 114
  living organisms *vs* machines 121
  as potential life criterion 79
  precellular 168
  reductive citric acid cycle
    variants 35
  of RNA 40
evolutionary transition 81,
  172, 173
exact accurate stoichiometry 32
exact chemistry 143
exact sciences 55–8, 67, 81
  exact theoretical science 170
exact theoretical biology 81, 115,
  147–52, 156
  chemotons as base units of 112
  life criteria in 181
  in philosophy 186
  resistance to 185
  role in averting catastrophe
    155–6
  units of 171
excitability
  of chemoton system 118
  as homeostasis mechanism 77
exobiology 80, 161

false models 185–6
feedback regulation 30, 123
  stoichiometric 21–2
flow of matter 177
fluid machines *see* automata,
  fluid
flux balance analysis 164–5
formaldehyde 41, 42
  in formose reaction 27, 28,
    34, 145
  and genesis of life 39, 146
  primordial 28–9

formose reaction 27–8, 34, 165
  as basis for primordial
    metabolism 28, 29
  and heredity 127
  and spontaneous genesis of
    life 39
Fox, S. and synthesis of living
  systems 144
free enthalpy content 94
function 68
functionalism 169, 171, 172, 173
fungal life 9

genes 13, 42, 52, 53
  molecular structure 15
  with properties of replicators
    172
genesis of life 11–12
  experiments 11
  spontaneous 39, 134, 141–2,
    146, 166
genetic alphabet 133
genetic circuits, theory of 58
genetic definition of life 159
genetic diversity 44
genetic homogeneity/uniqueness/
  autonomy 8
genetic material 83, 84, 130, 132
  double-stranded structure 131
  eukaryotic cells 9
  and hereditary properties/
    changes 125–6, 127
  primordial 39
  prokaryotic 6, 7
genetic regulation,
  Lwoff–Jacob–Monod model
    of 57
genetic texts 42
genetics, laws of 124–32
genotypic replicators 165
geometric limitations of
  membranes 33, 34
geometric order in automata 3
geometrical structure 8, 9
geometrical systems 64
geometry, exactness of 56
glycerol aldehyde 28, 29, 39
growing systems/growth 78–9,
  95–101

Haldane, J.B.S. on recipes for
  life 141
helical structure, deduction
  of 131, 132

hereditary changes   125
  in chemotons   114
  irreversibility/delayed reactions
    of   153–4
  mutational   128
  non-macromolecular   128
  non-mutational   127
  as potential life criterion   79
  in spherules   106
  see also mutation(s)
hereditary information   14, 17–18
hereditary optimization/
    adaption   44
hereditary properties
  changeable   42–3
  of chemotons   114
  and genetic information   125–6
  immutable   42, 44
  without hereditary changes   35
hereditary substance   107
heredity   35, 125
  limited   107
heterotrophs   168
holistic replicators   127, 133, 165
homeorhesis   77
homeostasis   73, 74, 77
  and cytoplasm   83, 99
  and inherent stability   77
  self-reproducing cyclic
      processes   100, 153
homogeneity, genetic   8
human life
  starting point   8–9
  supra-individual systems   10
hydrogen cyanide   41, 42
hypercycle   157

imidazolides   40, 41
individuality, criteria of   8
information   107
information-storage systems   85,
    106, 107, 131, 153
  chemical   39
  chemoton   107, 112–14, 119
  eukaryotic cells   9
  experimental validation   138–9
  as life criterion   77–8
  primitive   157
inherent units   176
inheritance   125, 130
  in paramecium   106
insect societies   10
  see also ant hills
interactors   171, 172
intragenomic conflict   163

irreversibility of events
    153–4, 155

jeewanus   145

kinetic events   115, 116
Krebs cycle   22, 23, 26, 87, 88–91,
    93, 96
  compared to autocatalytic
      cycles   96, 98
  general formula   91–3
  as soft system performing
      continuous work   90–1

learning   151
Leibniz, G.W.   12–13, 14
letter sequences   45–6, 47, 48
life
  abstract theoretical model
      systems   112
  cessation of, without death   79, 80
  definitions of   160–2
    Gánti's view   162
    Schrödinger's view   2
  levels of   1, 2, 81
    animal   7–9
    prokaryotic   6–7, 8
  origin of   132–40
    experiments on   132–3, 182
    see also biogenesis
  phenomena of   75–6
  recipes for   140–1
  spontaneous   39, 134, 141–2,
      146, 166
  see also primary life; secondary life
life criteria   74–80, 158–60
  and chemoton model   186
  chemotons satisfying   5, 112,
      113–14, 162, 179
  and engineering models   180–2
  selection/formulation/
      refinement   75, 159
  and units of exact theoretical
      biology   171
  weaknesses   181
  see also absolute life criteria;
      potential life criteria
light reaction   98
lipid vesicles, self-reproducing   140
liquid machines *see* automata, fluid
lithopanspermia   133
living state   75

living systems   59–61, 80, 158–9
  artificial production of   143
  chemical control   2–3
  chemical work in   71
  and chemotons   38, 143
  compensation for external
      change   74
  as dynamic functioning
      systems   72, 82
  equilibrium, preservation of   152
  exact definition   60
  incorporation of elements   98
  inherently individual   76
  internal constancy   73–4
  material composition   133
  minimal models   181–2
  non-enzymatic   136
  reversible/irreversible events   153
  self-reproducing nature   25, 33
  spontaneous   39, 134, 141–2,
      146, 166
  subsystems   83
  synthesis   140–7
    historical recipes   140–1
    total synthesis   143, 145
  vs non-living systems   133–4
  vs units of evolution   159
loop formation/coupling *see* RNA,
    abiotic loops
Lwoff–Jacob–Monod model of
    genetic regulation   57
Lyapunov functions   73

malic acid cycle   96–8, 100
material balance equations,
    stoichiometric   36
mathematics
  material balance equations   36
  and natural numbers   55
  plotting changes   20
  set theory   63
matter, organization of   1
mechanical systems   2–3
  equilibria   71
  periodicity in   72
mechanics   56
membranes   162
  artificial   102
  boundary, prokaryotic   6, 7
  cell   83, 84, 101
    geometric limitations   33
    double-molecule layers   34, 102
    fluid two-dimensional   4, 41, 101,
        102, 103
    experimental   138

formation/growth   103, 113
  and microsphere division
    34, 35
  genetic   35, 36
  molecular   33, 34
  phospholipid   102
  prebiotic   41
  raw materials from formose
    reaction   29
  selective permeability of
    84, 144
  spherule   105
  spontaneous formation
    33, 146
metabolic maps   38, 39
  see also connection diagrams
metabolic networks, primordial   28
metabolic processes   112
metabolic replicators   166–7
metabolic subsystem   4
  lack of in protocells   168
  lack of in viruses   7
metabolism   7, 16, 23, 28, 40, 48, 50,
    51, 53, 106, 115, 118, 119, 126,
    150, 167, 173, 177, 178
  Gánti's view   75
  as life criterion   76
  non-enzymatic   137, 168
metazoa   8
meteorite-delivered lipids   138
micelles, self-reproducing   168
microspheres   33
  division   34–5, 44
  lack of program control   35
  proteinoid   144, 157
minimal cycles, self-reproducing
    110–11
minimal systems   1, 3–6, 67, 81
  autocatalytic   103–4
  biological   111
  of cell   85
  chemotons as   114
  of cytoplasm   101
  and historical sequence   134
  RNA population as   161
minimum life   162–3
mitochondria   60, 61, 101
model systems
  biological   57
  in exact sciences   55
  of exact sciences   81
  idealization/abstraction   55, 56
  minimal
    of chemotons   4
    of living system   181–2
modular organisms   68
modular replication   165

molecular aggregates   102
molecular biology   58
molecules
  as basic units of chemistry   67
  as signs   108
monomer pair formation   132
monomers
  in chemical automata   30, 31, 32
  in chemotons   38
monomolecular layers   101
mortality   79–80
  of chemotons   114
multicellular organisms from
    non-living matter   149
multicellular organisms, organization
    of   10
multiplication   78–9
mutation(s)   14, 128
  in hypercycle   158
  mechanism   53
  in RNA   50
  in spherules   106
mutation rates   53, 133
mutational load   53, 157
mutative chemotons   114, 129
  sign sequence information
    storage   148
mutative molecular structures
    132
mycoplasmas   6, 37

NADH$_2$ 98
NASA Exobiology Program   161
negentropy   15
neuronal chemotons   151
non-living state   2, 75, 79, 80
nuclei   9
nucleic acid(s)
  bias   184
  failure of abiotic synthesis   140
  structure   145
  template polymerization
    31, 145, 147
nucleotides
  in RNA 45
  template polymerization   139
numeral signs   107–8
nutrient solution   104–5
nutrients   18, 37, 105

organisms as unitary systems   68
organization *vs* structure   177
organs, models of   150

origin of life *see* biogenesis; life,
    origin of
oscillatory reactions/systems   16, 23,
    25, 123–4
  in living systems   124
  predator–prey relationship   27
osmotic pressure and microsphere
    division   34–5
oxaloacetic acid
  in Krebs cycle   87, 89, 90, 91, 92
  in malic acid cycle   96–7

paradox of specificity   167
parasitic DNA   13
Parcelus' recipe for life   140–1
Pavlov, I. on life   55
periodic change   40
periodic regulation   16
periodicity   72
perpetuum mobile   87, 89
phenomena of life   75–6
phenotypic replicators   165
philosophy and biology   169–75
phospholipid formation   138
phosphorylation of adenine   96
photosynthesis   17–22, 98,
    138, 146
phylogenesis, irreversibility of   153
plant life   9
plastids   61
polymerization, program control
    in   30
polymers
  long, on clay   139
  as signs   108
potential life criteria   76, 78–81, 82,
    114, 177
  satisfied by chemotons   111
  satisfied by living systems   159–60
prebiotic conditions, simulations
    of   40
prebiotic evolution   136
primary life   8, 158
primordial conditions   133
primordial Earth conditions   134
  atmospheric   42, 136–7
  chemical reactions   137, 138
primordial genetic material   39
primordial pizza/crêpes scenario   135
primordial soup theory   134, 135
primordial water   135, 136
program control   30, 124
  'siphon' analogy   30, 31, 32
program-controlled automata
    30–1, 32

program-controlled
    self-reproduction   5
prokaryotes   6–7, 8, 61
proliferating chemical
    supersystems   104
proliferation   26, 33, 35
    rates   44
protein networks, reflexively
    autocatalytic   167
proteinoids   144, 157
protocells   168
purine bases   40, 41
pV_n storage macromolecule   119
pyrimidine bases   13, 40, 41
pyrite   40–1, 135

reaction chains/networks
    123, 137, 183
reaction kinetics   115
reaction rates   117–18, 127, 183
    and diversity   44
    and environmental changes   100
    stability   119
reaction specificities   183
reading mechanisms   107, 109
real life criteria *see* absolute life
    criteria
realization of chemotons   38–9, 41
receptor proteins   35–6
reductionism   57, 178, 184
reductive citric acid cycle
    26, 34, 35
regeneration of animals   59
regulation
    in living systems   78, 114
    periodic *see* periodic regulation
regulation technology   123
replication of information   78
replicators   127, 133, 166–7,
    172–3
    classification   165
    as significant units of
      selection   171
    of unlimited potential   133
reproduction   3, 4, 79
    see also autocatalysis
ribose   40
    in formose reaction   29
    and spontaneous genesis of
      life   39, 146
ribozymes   167–8
RNA
    abiotic loops
      formation/combination   47, 48,
      49, 51

    synthesizing enzyme
      assembly   52
    base sequences   43, 44, 45
    catalytic   167, 172, 175
    clover-leaf shape   48
    distribution on division   43
    enzymatic *see* ribozymes
    as enzyme   50–1, 52–3
    evolution   29, 40
    as living molecule   11
    molecules, nature of   45
    mutations   50
    nucleotides, building on of
      46–7
    prebiotic   39
      length   43
    single strand functions   53
    strand separation   40, 46, 52, 53
    synthesis   41, 43, 46, 47
    as template polymerization
      system   109
RNA macromolecules   48, 49
RNA populations as minimal
    systems   161
RNA world   161, 167

Santelices, B. on individuality   8
Schrödinger, E.   2
    on hereditary information   14
    search for genesis of life   12
science, victims of   152
secondary life   8
    in eukaryotic cells   9
    in plants/fungi   9
selection   50
    natural   35, 159, 171, 180
selective permeability of
    membranes   84, 144
self-organization, biological   176
self-reproducing mechanical
    devices   25
self-reproducing processes
    26, 29
    cycles   28, 96
      as basic unit of life
        subsystem   100
      in chemoton model   179
      malic acid cycle   97
      reversibility   113
self-reproduction   175
    and stability   100
    without information-carrying
      subsystem   125
semantics   57, 107
senescence   80

sensors   121–2
    effect on system   122–3
set theory   63
sets   63–8
side-reaction problem   166–7
signs   107–9, 123
    quantity/quality   107–8
    required by chemoton   132
    sequences for information
      storage   148
'six million dollar man'   143
social problems   1
solar radiation   39
souls   14
space, interstellar   134–5
spatial proximity in cells   84
spatial separation   32, 33, 85
specific operation   16
spherules   33, 103, 104–6, 107, 138
    division   105
    with double-molecular layer
      membranes   103
    molybdenum-containing   145
    self-reproducing   107, 110, 113,
      125, 161
      adaptation in   127
      as chemical automata   124
      and inheritance   106, 125
    two-cycle   126–8
splitting enzymes   50
spontaneous generation   39, 134,
    141–2, 146, 166
stability   73, 86
    inherent   76–7, 94, 112
    vs thermodynamic   87
    internal   82
    of simple chemical cycles
      99–100
stability criteria   73
state fields   20–1, 22
state–space approach to models   20
state spaces   23
state variables   153
steady/stationary state   73
stochastic corrector model
    163, 164
stoichiometric coupling   162,
    182–3
stoichiometric equations   116
stoichiometric feedback   21–2
stoichiometric state space   23
stoichiometry   4, 23, 177, 184–5
    chemical   21
structural inheritance system
    106
structuralism   169, 171, 172
    limitations   173

subsystems
  connected functioning 83
  of living cells 81–5
  three subsystem structure 145
  see also chemoton(s),
      subsystems
sugar phosphate cycle 34, 146
superorganisms 10
supra-individual systems 10
surface metabolism 135, 167
synthesizing enzymes 52
system–subsystem relations 82
system theory 63–4, 178,
      183, 184
systems 63–8
  definition 64
  divisible/unitary systems 66
Szathmáry, E. on prebiotic
      chemotons 42
Szent-Györgyi–Krebs cycle *see*
      Krebs cycle

telos/teleological assumptions 180
template molecules 44, 109
  with enzymatic capabilities 41
  length 157, 162
template polymerization 31–2, 109,
      140
  in chemotons 38, 44
  errors 129
  irreversibility 113–14
  long *vs* short templates 32
  of nucleic acids/nucleotides
      139, 145
  in primordial conditions 40
  synchronization 140
temporal synchrony of chemical
      reactions 145

tertiary life 10
theoretical biology 57, 170
  philosophical approach 169
  resistance to 185
  see also exact theoretical biology
theory reductionism 57
thermoplasmas 6–7, 37
three subsystem structure 8, 125
  chemotons 140
  compatibility/connection of
      subsystems 145
  prokaryotes 6, 7
'ticking' fluid automata 24, 26
time standards 123
tissue culture 59, 60
total synthesis 143, 147, 149
  of biology 149
  of living systems 143, 145
transfer RNA 48, 49, 157
transformation of energy 18
transforming agents 94
tricarboxylic acid cycle *see*
      Szent-Györgyi–Krebs cycle

uniqueness, genetic 8
unitary systems 66
units
  of biology 59
    philosophy of 173
    at prokaryotic level 61–2
  characterization 180
  of chemistry 58
    molecules as 67
  of crystallography 58–9, 67
  of evolution 159, 160,
      169, 170
    higher-level 163
    philosophy of 177–8, 180

reductive citric acid cycle
  variants 35
  search for 179
of exact theoretical biology 171
inherent 176
of life 58–63, 159, 160
  chemoton as 111, 114
  levels of organization 61
of selection 114, 127, 170, 171
systems 81
units analysis 169
unity 176
urea 41, 42

Van Helmont's recipe for life 141
variability 35
variation propagation in
      spherules 106
viruses 7, 60, 158, 159, 180
volume:surface area ratio 34
von Bertalanffy, L. 64
von Kiedrowski, G. on template
      length 32
von Neumann, J. on self-reproducing
      automata 124, 125

Watson–Crick model of DNA
      14–15, 45, 57, 124
work 18, 69, 71
work performance
  chemical motor of cell 90
  energetic conditions 68–9, 70
  see also continuous work
      performance
work-performing systems
      72, 74, 82